Three Classes of Nonlinear Stochastic Partial Differential Equations

Three Classes of Nonlinear Stochastic Partial Differential Equations

Jie Xiong

University of Macau, China &
The University of Tennessee, Knoxville, USA

World Scientific

NEW JERSEY • LONDON • SINGAPORE • BEIJING • SHANGHAI • HONG KONG • TAIPEI • CHENNAI

Published by

World Scientific Publishing Co. Pte. Ltd.
5 Toh Tuck Link, Singapore 596224
USA office: 27 Warren Street, Suite 401-402, Hackensack, NJ 07601
UK office: 57 Shelton Street, Covent Garden, London WC2H 9HE

British Library Cataloguing-in-Publication Data
A catalogue record for this book is available from the British Library.

THREE CLASSES OF NONLINEAR STOCHASTIC PARTIAL DIFFERENTIAL EQUATIONS

ISBN 978-981-4452-35-9

Printed in Singapore by World Scientific Printers.

在我的母亲卢祥云82岁生日之际

仅以此书献给她

Dedicate to my mother

Xiangyun Lu

on the occasion of her 82nd birthday

Preface

Superprocesses, which are also called Dawson-Watanabe processes, have been studied by many authors since the pioneer work of Dawson (1993) and Watanabe (1968). Much of the successes of these studies is attributed to the rich independent structures of the branching particles in the system leading to these processes. It is well-recognized of the need to introduce interaction among individual particles in the system. In this direction, Perkins' (2000) book provides an excellent introduction to the state of this topic of research.

Another possible extension is to introduce random environment applying to the whole system. This can be regarded as an interaction between individual particles and the environment in which the system lives. In this direction, two models were introduced: Wang (1998) and Skoulakis and Adler (2001) studied the limit of the branching particle systems where the motions of the particles are governed by the random environment. In this book, we will study these two models in a unified manner. We will also observe that the Zakai equation in the stochastic filtering theory is a special case of this general model. The idea of approximating optimal filter by branching particle system is well developed in the field of stochastic filtering (cf. Xiong (2008)). Mytnik (1996) considers the model when the environment affects the branching rate of the particles.

Since the aforementioned works, the study of measure-valued processes in a random environment has seen intensive research activity in recent years. Through these activities, some interesting nonlinear stochastic partial differential equations (SPDEs) have been derived. Because of the nonlinearity and non-Lipschitz continuity of their coefficients, the study of these SPDEs demands new tools. New techniques and concepts have been developed recently. They include the conditional Laplace transform technique, the conditional mild solution, and the bridge between SPDEs and some type of backward stochastic differential equations. The aim of this book is to provide an introduction to these topics and to attract more researchers to this exciting area of research.

Stochastic partial differential equations (SPDEs) is an important field in current research. We refer the reader to the books of Da Prato and Tubaro (2002), Da Prato and Zabczyk (1992), Kallianpur and Xiong (1995), Rozovskii (1990) for an introduction to this topic. Many authors have studied linear SPDEs. Here we only mention two recent papers: Gyöngy (2002) and Krylov (1999). Fine properties of solutions have been established and nonlinear SPDEs have also been studied. In addition, we mention two papers of Kotelenez (1992) and (1995) which are the closest to the present setting. In his case, the derivative of the solution is not involved in the noise term. To the best of our knowledge, the stochastic log-Laplace equation (5.1.4) does not fit into the setup of existing theory of SPDEs.

Although some of the results hold when the underlying space is of higher dimensions, the majority of them, either the conclusion or the method we use, hold for one-dimensional space only. Therefore, to be uniform throughout this monograph, we restrict ourselves to one-dimensional space.

Finally, I would like to acknowledge the help I received in preparing this manuscript. This project started at the University of Tennessee and was finished at the University of Macau. I would like to thank both institutions for their support. Financial support was also provided partially by DMS-0906907 of the National Science Foundation of the United States and by SRG022-FST12-XJ of the University of Macau. In addition, I would like to thank my student, Parisa Fatheddin, who read the whole manuscript, pointed out grammar errors and typos and gave many suggestions on how to improve the monograph. I would like to thank Zenghu Li, Thomas Rippl, Shanjian Tang, Xu Yang and Xiaowen Zhou for helpful suggestions. I am grateful to the staff of World Scientific Publishing Company for their patience and cooperation.

Jie Xiong
University of Macau, China &
The University of Tennessee, Knoxville, USA

Contents

Chapter 1

Introduction to Superprocesses

The study of measure-valued processes, also called superprocesses or Dawson-Watanabe processes, can be traced back to the works of Jirina (1958), (1964) and Watanabe (1968). Their approach was to consider superprocesses as the limits of branching particle systems arising from population models. Dawson (1975), followed by many others, later studied these processes systematically. In this chapter, we give a brief introduction to this topic and present some facts which are relevant to the stochastic partial differential equations (SPDEs) we shall study in this book.

1.1 Branching particle system

We introduce in this section a branching particle system arising from a population model. Let K_n be the number of particles at time 0 (i.e., the first generation particles), spatially distributed in $\mathbb{R}$ at points $x_1^n, x_2^n, \cdots, x_{K_n}^n$. Define the deterministic initial atomic measure as

$$\nu^n = \frac{1}{n} \sum_{i=1}^{K_n} \delta_{x_i^n},$$

where δ_x is the Dirac measure at x.

Let $\gamma > 0$ be a constant called the branching rate of the particle system. For the simplicity of notation, we assume that the lifespan of each particle in the system is $\frac{1}{n\gamma}$ (deterministic). In fact, the results which we will derive will not change if we assume that the particles have independent identically distributed lifespan with common exponential distribution with parameter $n\gamma$.

We begin by describing the branching mechanism of the particle system. The motions of the particles during their lifetimes will be described later.

At its death time, each particle gives birth to either zero or two particles with probability $\frac{1}{2}$ each. The daughters will take their mother's pre-death position and move on (again, to be described later) until their deaths at time $\frac{2}{n\gamma}$. This procedure will continue until all particles are dead. Note that the birth place of the particle coincides with her mother's death place.

Denote the collection of all multi-indices by $\mathcal{I}$, i.e.,

$$\mathcal{I} = \{\alpha = (\alpha_0, \alpha_1, \cdots, \alpha_N) : \ N \geq 0, \ \alpha_0 \in \mathbb{N}, \ \alpha_i \in \{1, 2\}, \ 1 \leq i \leq N\}.$$

We shall use the multi-index $\alpha \in I$ to denote a particle in this system. For example, $\alpha = (3, 1)$ stands for the older daughter of the third individual in the first generation. Denote the number of offspring of the particle α by $\alpha_{\#}$. From the definition of the branching system, we know that $\alpha_{\#}$ equals 0 or 2 with equal probability. Denote by $\alpha_b = \frac{N}{n\gamma}$ and $\alpha_d = \frac{N+1}{n\gamma}$ the birth and death times of the particle $\alpha = (\alpha_0, \alpha_1, \cdots, \alpha_N)$. The notation $\alpha \sim_n t$ means that the particle α is alive at time t, i.e.,

$$\alpha_b \leq t < \alpha_d.$$

Let $\{B_\alpha(t) : \alpha \in I\}$ be a family of independent d-dimensional Brownian motions. During its lifetime, the particle's motion is modeled by the Brownian motion $\xi_\alpha(t)$, $\alpha_b \leq t \leq \alpha_d$, given by

$$\xi_\alpha(t) = \xi_{\alpha-1}(\alpha_b) + B_\alpha(t) - B_\alpha(\alpha_b), \quad \alpha_b \leq t \leq \alpha_d,$$

where $\alpha - 1$ represents the mother of the particle α. Namely, if $\alpha = (\alpha_0, \alpha_1, \cdots, \alpha_N)$, then $\alpha - 1 = (\alpha_0, \alpha_1, \cdots, \alpha_{N-1})$.

We define the empirical measure process of the system as

$$X_t^n = \frac{1}{n} \sum_{\alpha \sim_n t} \delta_{\xi_\alpha(t)}. \tag{1.1.1}$$

Denote by $\mathcal{M}_F(\mathbb{R})$ the collection of all finite Borel measures on $\mathbb{R}$. We endow the space $\mathcal{M}_F(\mathbb{R})$ with the weak convergence topology, i.e., for $\mu_n, \mu \in \mathcal{M}_F(\mathbb{R})$, we say that $\mu_n \Rightarrow \mu$ if $\langle \mu_n, f \rangle \to \langle \mu, f \rangle$ for any $f \in C_b(\mathbb{R})$, where $\langle \mu, f \rangle$ denotes the integral of function f with respect to measure μ and $C_b(\mathbb{R})$ is the space of all bounded continuous functions on $\mathbb{R}$.

Next, we define the metric $d(\cdot, \cdot)$ on $\mathcal{M}_F(\mathbb{R})$. Let $f_0 = 1$ and $\{f_n\}_{n \geq 0}$ be a countable set whose linear span is dense in $C_b(\mathbb{R})$. For any $\mu, \nu \in \mathcal{M}_F(\mathbb{R})$, we define

$$d(\mu, \nu) = \sum_{n=0}^{\infty} 2^{-n} \left(|\langle \mu - \nu, f_n \rangle| \wedge 1 \right).$$

Then, $\mathcal{M}_F(\mathbb{R})$ is a Polish space whose topology is equivalent to that defined by the weak convergence of measures. The collection of all mappings from $\mathbb{R}_+$ to $\mathcal{M}_F(\mathbb{R})$ which are right continuous with left limit is denoted by $D([0,\infty),\mathcal{M}_F(\mathbb{R}))$. The space $D([0,\infty),\mathcal{M}_F(\mathbb{R}))$ is endowed with the Skorohod topology (cf. Ethier and Kurtz (1986) for definition).

Next, we proceed to proving the tightness of the family of measure-valued processes $\{X^n\}$ in $D([0,\infty),\mathcal{M}_F(\mathbb{R}))$.

Let $C_b^2(\mathbb{R})$ be the collection of real functions which are bounded with bounded derivatives up to order 2. For every $f \in C_b^2(\mathbb{R})$, we apply Itô's formula to $f(\xi_\alpha(t))$ to obtain

$$df(\xi_\alpha(t)) = \frac{1}{2}f''(\xi_\alpha(t))dt + f'(\xi_\alpha(t))dB_\alpha(t).$$

Throughout this book, we shall use f' to denote the derivative when f is a single variable function. Summing up over all α alive at time t, we get

$$d\langle X_t^n, f\rangle = \left\langle X_t^n, \frac{1}{2}f''\right\rangle dt + \frac{1}{n}\sum_{\alpha\sim_n t} f'(\xi_\alpha(t))dB_\alpha(t).$$

At a jumping time t, we have

$$\langle X_t^n - X_{t-}^n, f\rangle = \frac{1}{n}\sum_{\alpha_d=t}(\alpha_\# - 1)f(\xi_\alpha(t)). \tag{1.1.2}$$

We obtain by summing up,

$$\begin{aligned}\langle X_t^n, f\rangle &= \langle \nu^n, f\rangle + \int_0^t \left\langle X_s^n, \frac{1}{2}f''\right\rangle ds \\ &\quad + \frac{1}{n}\int_0^t \sum_{\alpha\sim_n s} f'(B_\alpha(s))dB_\alpha(s) \\ &\quad + \frac{1}{n}\sum_{s\le t}\sum_{\alpha\in I:\ \alpha_d=s}(\alpha_\# - 1)f(B_\alpha(s)) \\ &\equiv \langle \nu^n, f\rangle + Y_t^n(f) + U_t^n(f) + V_t^n(f). \end{aligned} \tag{1.1.3}$$

Note that $U_t^n(f)$ and $V_t^n(t)$ are two uncorrelated martingales.

Lemma 1.1.1. *Suppose that*

$$\sup_n \int_{\mathbb{R}} (1+|x|)\nu^n(dx) < \infty.$$

Then, $\{X^n\}$ satisfies the compact containment condition of Ethier and Kurtz (1986), i.e., for every $\eta > 0$ and $T > 0$ there is a compact set $\Gamma_{\eta,T} \subset \mathcal{M}_F(\mathbb{R})$ such that for any $t \in [0,T]$ and $n \in \mathbb{N}$,

$$P(X_t^n \in \Gamma_{\eta,T}) \ge 1-\eta.$$

Proof. By (1.1.3), we see that

$$\mathbb{E}\langle X_t^n, f\rangle = \langle \nu^n, f\rangle + \int_0^t \mathbb{E}\left\langle X_s^n, \frac{1}{2}f''\right\rangle ds.$$

Thus,

$$\mathbb{E}\langle X_t^n, f\rangle = \langle \nu^n, T_t f\rangle, \tag{1.1.4}$$

where T_t is the semigroup generated by $\frac{1}{2}\Delta$. Denote by $C_0^2(\mathbb{R})$ the collection of all functions on $\mathbb{R}$ which have compact supports and continuous derivatives up to order 2. Let $\{f_m\} \in C_0^2(\mathbb{R})$ be an increasing sequence with limit (as $m \to \infty$) $1 + |x|$. Replacing f in (1.1.4) by f_m and taking $m \to \infty$, we obtain

$$\begin{aligned}\mathbb{E}\int_{\mathbb{R}}(1+|x|)X_t^n(dx) &= \int_{\mathbb{R}} \nu^n(dx)\mathbb{E}_x(1+|B(t)|)\\ &\le \int_{\mathbb{R}}\left(1+|x|+\sqrt{t}\right)\nu^n(dx) \le K_1,\end{aligned}$$

where $B(t)$ is a Brownian motion and K_1 is a constant. Note that, in the last inequality, we used the fact that $B(t)$ is a normal random variable with mean x and variance t, denoted as $B(t) \sim N(x,t)$.

Let

$$\Gamma_{\eta,T} = \left\{\mu \in \mathcal{M}_F(\mathbb{R}) : \int_{\mathbb{R}}(1+|x|)\mu(dx) \le K_1\eta^{-1}\right\}.$$

Then, $\Gamma_{\eta,T}$ is compact in $\mathcal{M}_F(\mathbb{R})$ and

$$P(X_t^n \notin \Gamma_{\eta,T}) \le K_1^{-1}\eta\mathbb{E}\int_{\mathbb{R}}(1+|x|)X_t^n(dx) \le \eta.$$

□

Now, we only need to prove that for any $f \in C_b^2(\mathbb{R})$, the family $\{\langle X^n, f\rangle\}$ is tight in $D([0,\infty),\mathbb{R})$. It is easy to see that

$$\begin{aligned}\langle U^n(f)\rangle_t &= \frac{1}{n^2}\int_0^t \sum_{\alpha\sim_n s} |f'(\xi_\alpha(s))|^2\, ds\\ &= \frac{1}{n}\int_0^t \left\langle X_s^n, |f'|^2\right\rangle ds.\end{aligned} \tag{1.1.5}$$

To calculate the quadratic variation process of $V^n(f)$, we note that

$$\begin{aligned}
&\mathbb{E}\left((V_t^n(f) - V_s^n(f))^2 | \mathcal{F}_s\right) \\
&= \frac{1}{n^2}\mathbb{E}\left(\left(\sum_{s<r\leq t}\sum_{\alpha_d=r}(\alpha_\# - 1)f(\xi_\alpha(r-))\right)^2 \middle| \mathcal{F}_s\right) \\
&= \frac{1}{n^2}\sum_{s<r\leq t}\sum_{\alpha_d=r}\mathbb{E}\left((\alpha_\# - 1)^2 f^2(\xi_\alpha(r-))\middle|\mathcal{F}_s\right) \\
&= \frac{1}{n^2}\sum_{s<r\leq t}\sum_{\alpha_d=r}\mathbb{E}\left(f^2(\xi_\alpha(r-))\middle|\mathcal{F}_s\right) \\
&= \mathbb{E}\left(\frac{1}{n}\sum_{s<r\leq t}\langle X_r^n, f^2\rangle \middle| \mathcal{F}_s\right) \\
&= \mathbb{E}\left(\int_s^t \langle X_r^n, \gamma f^2\rangle\, dr \middle| \mathcal{F}_s\right).
\end{aligned}$$

Thus,

$$\langle V^n(f)\rangle_t = \int_0^t \left\langle X_r^n, \gamma f^2\right\rangle dr. \tag{1.1.6}$$

Next, we apply the result to $f = 1$. In this case,

$$\langle X_t^n, 1\rangle = \langle \nu^n, 1\rangle + V^n(1).$$

Then,

$$\begin{aligned}
\mathbb{E}\sup_{s\leq t}\langle X_s^n, 1\rangle^2 &\leq 2\mathbb{E}\langle \nu^n, 1\rangle^2 + 8\gamma\mathbb{E}\int_0^t \langle X_r^n, 1\rangle\, dr \\
&= 2\mathbb{E}\langle \nu^n, 1\rangle^2 + 8\gamma t\mathbb{E}\langle \nu^n, 1\rangle \\
&\leq K.
\end{aligned} \tag{1.1.7}$$

Now, we fix $f \in C_b^2(\mathbb{R})$. As

$$\langle V^n(f) + U^n(f)\rangle_t = \langle V^n(f)\rangle_t + \langle U^n(f)\rangle_t,$$

it follows from (1.1.5), (1.1.6) and (1.1.7) that $\{\langle V^n(f) + U^n(f)\rangle\}$ is $\mathbb{C}$-tight. Similarly, $\{Y^n(f)\}$ is also $\mathbb{C}$-tight. By Theorem 6.1.1 in Kallianpur and Xiong (1995), we see that $\{\langle X^n, f\rangle\}$ is tight in $D([0,\infty),\mathbb{R})$. Therefore, $\{X^n\}$ is tight in $D([0,\infty),\mathcal{M}_F(\mathbb{R}^d))$.

We are now ready to characterize the limit of $\{X^n\}$.

Theorem 1.1.2. *Assume that $\nu^n \Rightarrow \nu$ in $\mathcal{M}_F(\mathbb{R})$ and*

$$\sup_n \int_{\mathbb{R}} \left(1 + |x|^2\right)\nu^n(dx) < \infty.$$

Then, the sequence $\{X^n\}$ *is tight in* $D([0,\infty),\mathcal{M}_F(\mathbb{R}))$. *Furthermore, the limit* X *lies in* $C(\mathbb{R}_+,\mathcal{M}_F(\mathbb{R}))$ *and is a solution of the following martingale problem (MP): For all* $f\in C_b^2(\mathbb{R})$,

$$M_t(f)\equiv\langle X_t,f\rangle-\langle\nu,f\rangle-\int_0^t\left\langle X_s,\frac{1}{2}f''\right\rangle ds \tag{1.1.8}$$

is a continuous square integrable martingale such that $M_0(f)=0$ *and*

$$\langle M(f)\rangle_t=\int_0^t\gamma\left\langle X_s,f^2\right\rangle ds. \tag{1.1.9}$$

Proof. The tightness has already been demonstrated above. Now we prove that X is a continuous solution to the martingale problem (1.1.8, 1.1.9). By (1.1.3), we see that

$$M_t^n(f)\equiv\langle X_t^n,f\rangle-\langle\nu^n,f\rangle-\int_0^t\left\langle X_s^n,\frac{1}{2}f''\right\rangle ds$$

is a martingale. Further, $M_t^n(f)\to M_t(f)$, and for $t\le T$,

$$|M_t^n(f)|\le K\left(1+\sup_{t\le T}\langle X_t^n,1\rangle\right).$$

By (1.1.7), we can pass to the limit to conclude that $M_t(f)$ is a martingale.

On the other hand,

$$N_t^n(f)\equiv M_t^n(f)^2-\langle U^n(f)\rangle_t-\langle V^n(f)\rangle_t$$

is a martingale which converges to

$$N_t(f)=M_t(f)^2-\int_0^t\gamma\left\langle X_s,f^2\right\rangle ds.$$

Similar to the previous paragraph, we may pass to the limit to conclude that $N_t(f)$ is a martingale. Therefore,

$$\langle M(f)\rangle_t=\int_0^t\gamma\left\langle X_s,f^2\right\rangle ds.$$

Namely, X_t is a solution to martingale problem (1.1.8, 1.1.9).

Finally, we prove the continuity of X. Applying (1.1.2), we have

$$\begin{aligned}\mathbb{E}\sum_{s\le t}\langle X_s^n-X_{s-}^n,f\rangle^2&=\mathbb{E}\sum_{s\le t}\left|\frac{1}{n}\sum_{\alpha_d=t}(\alpha_\#-1)f(\xi_\alpha(t-))\right|^2\\&\le\frac{K}{n^2}\mathbb{E}\sum_{s\le t}\langle X_s^n,1\rangle^2\\&\le\frac{K}{n}\to 0.\end{aligned}$$

Using Fatou's lemma, we get

$$\mathbb{E}\sum_{s\le t}\langle X_s - X_{s-}, f\rangle^2 = 0,$$

which implies the continuity of X. □

The uniqueness of the solution to the MP will be treated in the next two sections by two different approaches.

1.2 The log-Laplace equation

Here we present the first method in proving the uniqueness of the solution to MP (1.1.8, 1.1.9). The partial differential equation (PDE), which we will use in this method, has many other applications in the study of various properties of the superprocesses. We will not present these applications here and instead, we refer the interested reader to the books of Dawson (1993), Etheridge (2000), Le Gall (1999), Li (2011) and Perkins (2002).

Let $C_b^+(\mathbb{R})$ be the collection of all non-negative bounded continuous functions on $\mathbb{R}$.

Theorem 1.2.1. *If X_t is a solution to MP (1.1.8, 1.1.9), then for any $f \in C_b^+(\mathbb{R})$, we have*

$$\mathbb{E}\exp\left(-\langle X_t, f\rangle\right) = \exp\left(-\langle \nu, u_t\rangle\right), \tag{1.2.1}$$

where u_t is the unique solution to the following PDE

$$\begin{cases} \frac{\partial}{\partial t}u_t(x) = \frac{1}{2}\Delta u_t(x) - \frac{\gamma}{2}u_t(x)^2 \\ u_0(x) \quad = f(x). \end{cases} \tag{1.2.2}$$

As a consequence, MP (1.1.8, 1.1.9) is well-posed.

Proof. Here we only give a sketch of the proof. Let t be fixed and $s \in [0,t]$ be the time variable. By Itô's formula, we have

$$\begin{aligned} d\langle X_s, u_{t-s}\rangle &= \left\langle X_s, \frac{1}{2}\Delta u_{t-s}\right\rangle ds + \langle dM_s, u_{t-s}\rangle \\ &\quad - \left\langle X_s, \frac{1}{2}\Delta u_{t-s} - \frac{1}{2}\gamma u_{t-s}^2\right\rangle ds \\ &= \left\langle X_s, \frac{1}{2}\gamma u_{t-s}^2\right\rangle ds + \langle dM_s, u_{t-s}\rangle . \end{aligned} \tag{1.2.3}$$

Applying Itô's formula to (1.2.3), we get

$$\begin{aligned}
& d\exp\left(-\left\langle X_s, u_{t-s}\right\rangle\right) \\
&= -\exp\left(-\left\langle X_s, u_{t-s}\right\rangle\right)\left(\left\langle X_s, \frac{1}{2}\gamma u_{t-s}^2\right\rangle ds + \left\langle dM_s, u_{t-s}\right\rangle\right) \\
&\quad + \frac{1}{2}\exp\left(-\left\langle X_s, u_{t-s}\right\rangle\right)\gamma\left\langle X_s, u_{t-s}^2\right\rangle ds \\
&= \exp\left(-\left\langle X_s, u_{t-s}\right\rangle\right)\left\langle dM_s, u_{t-s}\right\rangle . \qquad (1.2.4)
\end{aligned}$$

Thus,

$$\exp\left(-\left\langle X_t, f\right\rangle\right) = \exp\left(-\left\langle \nu, u_t\right\rangle\right) + \int_0^t \exp\left(-\left\langle X_s, u_{t-s}\right\rangle\right)\left\langle dM_s, u_{t-s}\right\rangle . \qquad (1.2.5)$$

Taking expectations on both sides of (1.2.5), we see that (1.2.1) holds. □

The PDE (1.2.2) is called the log-Laplace equation for superprocess X_t.

Note that the above theorem only works when the underlying motion for each particle is a time-homogeneous Markov process. For the time-inhomogeneous case, e.g., when the underlying motion processes are replaced by independent identically distributed time-inhomogeneous Markov processes, the Laplace transform needs to be formulated using two time-parameters. We reformulate the theorem above so that it can be generalized to the time-inhomogeneous case. This point of view is crucial when we study the superprocesses in a random environment (SPRE).

Theorem 1.2.2. *If X_t is a solution to MP (1.1.8, 1.1.9), then for any $f \in C_b^+(\mathbb{R})$, we have*

$$\mathbb{E}\exp\left(-\left\langle X_t, f\right\rangle\right) = \exp\left(-\left\langle \nu, v_{0,t}\right\rangle\right), \qquad (1.2.6)$$

where, for fixed t, $\{v_{s,t} : s \in [0,t]\}$ is the unique solution to the following backward PDE

$$\begin{cases} \frac{\partial}{\partial s} v_{s,t}(x) = -\frac{1}{2}\Delta v_{s,t}(x) + \frac{\gamma}{2} v_{s,t}(x)^2 \\ v_{t,t}(x) \quad = f(x). \end{cases} \qquad (1.2.7)$$

1.3 The moment duality

In this section, we present another method in proving the uniqueness for the solution to MP (1.1.8, 1.1.9). We begin by stating the following duality result taken from Ethier and Kurtz (1986).

For a stochastic process X_t, we denote by $\mathcal{F}_t^X = \sigma(X_s : s \le t)$ the σ-field generated by X up to time t.

Theorem 1.3.1. *Let E_1 and E_2 be two Polish spaces and suppose X_t and Y_t are two independent processes taking values in E_1 and E_2, respectively. Let f, g, h be real functions on $E_1 \times E_2$, and α and β are real functions defined on E_1 and E_2, respectively. If*

i) for any $y \in E_2$,

$$f(X_t, y) - \int_0^t g(X_s, y)ds$$

is an $\mathcal{F}_t^X$-martingale;

ii) for any $x \in E_1$,

$$f(x, Y_t) - \int_0^t h(x, Y_s)ds$$

is an $\mathcal{F}_t^Y$-martingale; and

iii)

$$g(x, y) + \alpha(x)f(x, y) = h(x, y) + \beta(y)f(x, y),$$

then,

$$\begin{aligned} &\mathbb{E}\left[f(X_t, Y_0)\exp\left(\int_0^t \alpha(X_s)ds\right)\right] \\ &= \mathbb{E}\left[f(X_0, Y_t)\exp\left(\int_0^t \beta(Y_s)ds\right)\right]. \end{aligned} \tag{1.3.1}$$

Proof. Let t be fixed. Then,

$$\begin{aligned} &\frac{d}{ds}\mathbb{E}\left[f(X_s, Y_{t-s})\exp\left(\int_0^s \alpha(X_r)dr + \int_0^{t-s} \beta(Y_r)dr\right)\right] \\ &= \mathbb{E}\Bigg[(g(X_s, Y_{t-s}) - h(X_s, Y_{t-s})) \\ &\qquad \times \exp\left(\int_0^s \alpha(X_r)dr + \int_0^{t-s} \beta(Y_r)dr\right)\Bigg] \\ &\quad + \mathbb{E}\Bigg[f(X_s, Y_{t-s})\exp\left(\int_0^s \alpha(X_r)dr + \int_0^{t-s} \beta(Y_r)dr\right) \\ &\qquad \times (\alpha(X_s) - \beta(Y_{t-s}))\Bigg] \\ &= 0. \end{aligned}$$

Identity (1.3.1) follows easily. □

Lemma 1.3.2. *Let X_t be a solution to MP (1.1.8, 1.1.9). Then, for any $f \in C_b^2(\mathbb{R}^n)$, the stochastic process*

$$F(X_t,(n,f)) - \int_0^t G(X_s,(n,f))ds \tag{1.3.2}$$

is a martingale, where

$$F(\mu,(n,f)) = \langle \mu^{\otimes n}, f\rangle \equiv \int_{\mathbb{R}} \cdots \int_{\mathbb{R}} f(x_1,\cdots,x_n)\mu(dx_1)\cdots\mu(dx_n)$$

and

$$G(\mu,(n,f)) = \gamma \sum_{j=1}^n \left\langle \mu^{\otimes n}, \frac{1}{2}\partial_j^2 f\right\rangle + \gamma \sum_{1\le j<k\le n} \left\langle \mu^{\otimes(n-1)}, \Phi_{jk} f\right\rangle,$$

while $\partial_j f$ stands for the partial derivative with respect to the jth variable x_j, and $\Phi_{jk} f$ is a function of $n-1$ variables with the original jth and kth variables coalesced, i.e.,

$$(\Phi_{jk}f)(x_1,\cdots,x_{n-1}) = f(x_1,\cdots,x_{k-1},x_j,x_k,\cdots,x_{n-1}).$$

We call Φ_{jk} the coalescent operator.

Proof. Using approximation if necessary, we may and will assume that

$$f(x_1,\cdots,x_n) = \prod_{j=1}^n f_j(x_j), \qquad f_j \in C_b^2(\mathbb{R}).$$

As

$$M_t(f+g) = M_t(f) + M_t(g),$$

by (1.1.9), it is easy to verify that

$$\langle M(f), M(g)\rangle_t = \gamma \int_0^t \langle X_s, fg\rangle\, ds.$$

Applying Itô's formula, we have

$$\begin{aligned} dF(X_t,(n,f)) &= \sum_{j=1}^n \left(\prod_{i\ne j}\langle X_t, f_i\rangle\right)\left\langle X_t, \frac{1}{2}f_j''\right\rangle dt \\ &\quad + \gamma \sum_{j<k}\left(\prod_{i\notin\{j,k\}} \langle X_t, f_i\rangle\right)\langle X_t, f_j f_k\rangle\, dt + d(mart.) \\ &= G(X_t,(n,f))dt + d(mart.), \end{aligned}$$

where the notation $d(mart.)$ stands for the differential of a martingale. This proves that (1.3.2) is a martingale. □

To define the moment dual of the superprocess, we now take $E_1 = \mathcal{M}_F(\mathbb{R})$ and

$$E_2 = \{(n, f) : n \in \mathbb{N} \text{ and } f \in C_b^2(\mathbb{R}^n)\}.$$

Let n_t be Kingman's coalescent process, i.e., it is an $\mathbb{N}$-valued Markov process with transition intensities $\{q_{i,j}\}$ given by

$$q_{i,i-1} = -q_{i,i} = \frac{\gamma i(i-1)}{2}$$

and $q_{i,j} = 0$ for all other pairs, where $\mathbb{N} = \{1, 2, \cdots\}$ denotes the collection of natural numbers. Let $\tau_0 = 0$, $\tau_{n_0} = \infty$ and $\{\tau_k : 1 \le k < n_0\}$ be the sequence of jump times of the process $\{n_t : t \ge 0\}$. Let $\{S_k : 1 \le k < n_0\}$ be a sequence of random operators such that

$$P\left(S_k = \Phi_{ij} | n_{\tau_k -} = \ell\right) = \frac{2}{\ell(\ell-1)}, \qquad 1 \le i < j \le \ell.$$

Namely, S_k picks up one of the coalesent operators at random. Recall that T_t is the semigroup generated by $\frac{1}{2}\Delta$. For $\tau_k \le t < \tau_{k+1}$, define

$$f_t = T_{t-\tau_k}^{\otimes n_{\tau_k}} S_k T_{\tau_k - \tau_{k-1}}^{\otimes n_{\tau_{k-1}}} S_{k-1} \cdots T_{\tau_2-\tau_1}^{\otimes n_{\tau_1}} S_1 T_{\tau_1}^{\otimes n_0} f_0,$$

where for any $f \in C_b(\mathbb{R}^n)$,

$$T_t^{\otimes n} f(x) = \int_{\mathbb{R}^n} f(y) \prod_{j=1}^n p_t(x_j - y_j) dy, \qquad \forall\, x \in \mathbb{R}^n,$$

and

$$p_t(x) = \frac{1}{\sqrt{2\pi}} e^{-\frac{1}{2}x^2}, \qquad \forall\, x \in \mathbb{R},$$

i.e., $T_t^{\otimes n}$ is the $C_b(\mathbb{R}^n)$-semigroup of the n-dimensional Brownian motion. Then, for any $\mu \in \mathcal{M}_F(\mathbb{R})$,

$$F(\mu, (n_t, f_t)) - \int_0^t H(\mu, (n_s, f_s)) ds \tag{1.3.3}$$

is a martingale, where

$$H(\mu, (n, f)) = G(\mu, (n, f)) + \frac{1}{2}\gamma n(n-1) F(\mu, (n, f)).$$

We will refer to T_t as the Brownian semigroup.

Theorem 1.3.3. *For any $f \in C_b(\mathbb{R}^n)$, we have*

$$\mathbb{E}F(X_t, (n, f)) = \mathbb{E}\left(F(\mu, (n_t, f_t)) \exp\left(\int_0^t \frac{\gamma}{2} n_s(n_s - 1) ds\right)\right). \tag{1.3.4}$$

Proof. The statement follows from Theorem 1.3.1 and the martingale properties of (1.3.2) and (1.3.3). □

As a corollary of the moment duality, we write out the first two moments explicitly. These formulas will be useful in Section 1.4 in deriving the existence of the density random field for the superprocess. Recall that $\nu \in \mathcal{M}_F(\mathbb{R})$ is the initial measure of the process X_t.

Corollary 1.3.4. *For any $t \geq 0$ and $\phi \in C_b(\mathbb{R})$, we have*

$$\mathbb{E}\langle X_t, \phi\rangle = \langle \nu, T_t\phi\rangle$$

and

$$\mathbb{E}\langle X_t, \phi\rangle^2 = \langle \nu, T_t\phi\rangle^2 + \gamma \int_0^t \int_{\mathbb{R}} \int_{\mathbb{R}} p_s(x-y)\left(T_{t-s}\phi(y)\right)^2 dy\nu(dx)ds.$$

Proof. Recall that $F(\nu,(n,f)) = \langle \nu^{\otimes n}, f\rangle$. To obtain the first equality, we set $n_0 = 1$ and $f_0 = \phi$. Then,

$$\mathbb{E}\langle X_t, \phi\rangle = \langle \nu, f_t\rangle = \langle \nu, T_t\phi\rangle .$$

Now, we set $n_0 = 2$ and $f_0 = \phi^{\otimes 2}$. By (1.3.4), we obtain

$$\begin{aligned}\mathbb{E}\langle X_t, \phi\rangle^2 &= \mathbb{E}\left(F(\nu,(n_t,f_t))1_{\tau_1>t}e^{\gamma t}\right) + \int_0^t \gamma e^{-\gamma s}e^{\gamma s}\left\langle \nu, T_{t-s}(f_s^2)\right\rangle ds \\ &= \langle \nu, T_t\phi\rangle^2 + \gamma \int_0^t \int_{\mathbb{R}} \int_{\mathbb{R}} p_s(x-y)\left(T_{t-s}\phi(y)\right)^2 dy\nu(dx)ds.\end{aligned}$$

□

In general, moments of a random variable do not determine the distribution of that random variable. Many sufficient conditions for the moments to determine the distributions are known in the literature. The following sufficient condition, whose proof can be found on page 342 in the book of Billingsley (1979), will be useful in this book.

Theorem 1.3.5. *Let μ be a probability measure on $\mathbb{R}$ having finite moments $\alpha_k = \int_{\mathbb{R}} x^k \mu(dx)$ of all orders. If the power series $\sum_{k=0}^{\infty} \frac{\alpha_k}{k!} r^k$ has a positive radius of convergence, then μ is the only probability measure with the moments α_k, $k \geq 1$.*

Corollary 1.3.6. *Martingale problem (1.1.8, 1.1.9) is well-posed.*

Proof. We only need to prove the uniqueness of the distribution for random variable $\langle X_t, \phi\rangle$ with $\phi \in C_b(\mathbb{R})$ and $t \geq 0$ fixed. Let $f = \phi^{\otimes n} \in C_b(\mathbb{R}^n)$. Then,

$$\alpha_n \equiv \mathbb{E}\left(\langle X_t, \phi\rangle^n\right) \leq \tilde{\alpha}_n \|\phi\|_\infty^n,$$

where $\tilde{\alpha}_n = \mathbb{E}\left(\langle X_t, 1\rangle^n\right)$ and

$$\|\phi\|_\infty = \sup\{|\phi(x)| : x \in \mathbb{R}\}$$

is the supremum norm of ϕ.

Let

$$\psi_t(\lambda) = \sum_{n=0}^{\infty} \frac{\tilde{\alpha}_n}{n!}\lambda^n = \mathbb{E}e^{\lambda\langle X_t, 1\rangle}.$$

By Theorem 1.2.1, for $\lambda < 0$, we have

$$\psi_t(\lambda) = \exp\left(\langle \mu, 1\rangle\, \phi_t(\lambda)\right),$$

where

$$\begin{cases} \partial_t \phi_t(\lambda) = \frac{\gamma}{2}\phi_t(\lambda)^2 \\ \phi_0(\lambda) = \lambda. \end{cases}$$

So,

$$\phi_t(\lambda) = \left(\lambda^{-1} - \frac{1}{2}\gamma t\right)^{-1}$$

is analytic on $\lambda \in \left(-\infty, 2(\gamma t)^{-1}\right)$. Thus, the power series $\sum_{n=0}^\infty \frac{\tilde{\alpha}_n}{n!}\lambda^n$ has a positive radius of convergence $\lambda_0 = 2(\gamma t)^{-1}$. Hence, the power series $\sum_{n=0}^\infty \frac{\alpha_n}{n!}\lambda^n$ has a positive radius of convergence $\lambda \geq \lambda_0/\|\phi\|_\infty > 0$. This implies uniqueness of the distribution of $\langle X_t, \phi\rangle$. □

1.4 The SPDE for the density

In this section, we will prove the existence of the density random field for our described superprocess and derive the SPDE satisfied by this random field. Furthermore, making use of the SPDE, we will prove the joint continuity of the random field in the spatial and time variables. For these purposes, the following estimate proves to be useful. In the next lemma and throughout the rest of this monograph, we will use the following convention on constants: We use K with a subscript consecutively and restart from 1 with each new theorem. For example, K_1 in different theorems might have different values. Sometimes, when there is no confusion possible from the context, we even drop the subscript and use K for a constant whose exact value can be changed from place to place.

Lemma 1.4.1. *For any $T > 0$ and $\alpha < \frac{1}{2}$, there exists a constant $K = K(\alpha, T)$ such that for any $t \leq T$ and δ, $\delta' > 0$,*

$$\int_0^t \int_{\mathbb{R}} |p_{s+\delta}(x) - p_{s+\delta'}(x)|^2 dx ds \leq K|\delta - \delta'|^\alpha.$$

Proof. Note that for any $0 \leq s < t \leq T$,

$$\begin{aligned} |p_s(y) - p_t(y)| &\leq \int_s^t \frac{1}{2} r^{-1} p_r(y) \left(1 + \frac{y^2}{r}\right) dr \\ &\leq K(T) s^{-1} |t - s| I_{s,t}(y), \end{aligned}$$

where $K(T)$ is a constant, that may depend on T, and

$$I_{s,t}(y) = p_s(y) + p_t(y) + p_{2s}(y) + p_{2t}(y).$$

Write

$$\begin{aligned} |p_{s+\delta}(x) - p_{s+\delta'}(x)|^2 &= |p_{s+\delta}(x) - p_{s+\delta'}(x)|^\alpha \\ &\quad \times |p_{s+\delta}(x) - p_{s+\delta'}(x)|^{2-\alpha}, \end{aligned}$$

and estimate the first factor by $K_1 s^{-\alpha} |\delta - \delta'|^\alpha I_{s+\delta, s+\delta'}(x)^\alpha$ and the second fact by $I_{s+\delta,s+\delta'}(x)^{2-\alpha}$, we have

$$\begin{aligned} &\int_0^t \int_{\mathbb{R}} |p_{s+\delta}(x) - p_{s+\delta'}(x)|^2 dx ds \\ &\leq K_1 \int_0^t \int_{\mathbb{R}} s^{-\alpha} |\delta - \delta'|^\alpha I_{s+\delta,s+\delta'}(x)^2 dx ds \\ &\leq K_1 \int_0^t s^{-\alpha} |\delta - \delta'|^\alpha s^{-1/2} ds \\ &\leq K|\delta - \delta'|^\alpha, \end{aligned}$$

where the last inequality follows from the fact that $\alpha + \frac{1}{2} < 1$. □

Let $H_0 = L^2(\mathbb{R})$ be the Hilbert space consisting of all square-integrable functions with Hilbertian norm $\|\cdot\|_0$ given by

$$\|f\|_0^2 = \int_{\mathbb{R}} f(x)^2 dx,$$

where the corresponding inner product is denoted by $\langle \cdot, \cdot \rangle_0$.

Theorem 1.4.2. *Suppose that ν has density $u_0 \in H_0$. Then, there exists a random field $u \in L^2(\Omega \times [0,T] \times \mathbb{R}, P(d\omega)dtdx)$ such that*

$$\lim_{\delta \to 0+} \mathbb{E} \int_0^T \int_{\mathbb{R}} |T_\delta X_t(x) - u_t(x)|^2 dx dt = 0. \tag{1.4.1}$$

As a consequence, X_t has density u_t with respect to the Lebesgue measure.

Proof. For any δ, $\delta' > 0$, by the moment duality for superprocess, we have

$$\begin{aligned}
&\mathbb{E}|T_\delta X_t(x) - T_{\delta'} X_t(x)|^2 \\
&= \mathbb{E}\left\langle X_t, p_\delta(x-\cdot) - p_{\delta'}(x-\cdot)\right\rangle^2 \\
&= \left\langle \nu, p_{t+\delta}(x-\cdot) - p_{t+\delta'}(x-\cdot)\right\rangle^2 \\
&\quad + \gamma \int_0^t \int_{\mathbb{R}} \int_{\mathbb{R}} p_s(z-y) \left(p_{t-s+\delta}(x-y) - p_{t-s+\delta'}(x-y)\right)^2 dy\nu(dz)ds.
\end{aligned}$$

Thus,

$$\mathbb{E}\int_0^T \int_{\mathbb{R}} |T_\delta X_t(x) - u_t(x)|^2 dxdt = I_1 + I_2, \tag{1.4.2}$$

where

$$I_1 = \int_0^T dt \int_{\mathbb{R}} dx \left\langle \nu, p_{t+\delta}(x-\cdot) - p_{t+\delta'}(x-\cdot)\right\rangle^2$$

and

$$\begin{aligned}
I_2 = \gamma \int_0^T dt \int_{\mathbb{R}} dx \int_0^t \int_{\mathbb{R}} \int_{\mathbb{R}} p_s(z-y) \\
\times \left(p_{t-s+\delta}(x-y) - p_{t-s+\delta'}(x-y)\right)^2 dy\nu(dz)ds.
\end{aligned}$$

Now, we estimate I_1 and I_2 as follows.

$$I_1 \le 2\int_0^T dt \int_{\mathbb{R}} dx \int_{\mathbb{R}} u_0(y)^2 |p_{t+\delta}(x-y) - p_{t+\delta'}(x-y)|dy.$$

Write

$$\begin{aligned}
|p_{t+\delta}(x-y) - p_{t+\delta'}(x-y)| = |p_{t+\delta}(x-y) - p_{t+\delta'}(x-y)|^\alpha \\
\times |p_{t+\delta}(x-y) - p_{t+\delta'}(x-y)|^{1-\alpha},
\end{aligned}$$

and estimate the first factor by $K_1 t^{-\alpha}|\delta - \delta'|^\alpha I_{t+\delta,t+\delta'}(y)^\alpha$ and the second factor by $I_{t+\delta,t+\delta'}(y)^{1-\alpha}$, we may continue with

$$\begin{aligned}
I_1 &\le K_2 \int_0^T dt \|u_0\|_0^2 \int_{\mathbb{R}} t^{-\alpha}|\delta - \delta'|^\alpha I_{t+\delta,t+\delta'}(y)dy \\
&\le K_3|\delta - \delta'|^\alpha,
\end{aligned}$$

where $\alpha \in (0,1)$ and K_3 is a constant which may depend on T, α, $\|u_0\|_0$.

On the other hand,

$$\begin{aligned}
I_2 &= \gamma \int_0^T dt \int_{\mathbb{R}} dy \int_{\mathbb{R}} \nu(dz) p_s(z-y) \\
&\quad \times \int_0^t \int_{\mathbb{R}} \left(p_{t-s+\delta}(x-y) - p_{t-s+\delta'}(x-y)\right)^2 dxds \\
&\le K_4 \int_0^T dt \int_{\mathbb{R}} dy \int_{\mathbb{R}} \nu(dz) p_s(z-y)|\delta - \delta'|^\alpha \\
&= K_4 T\nu(\mathbb{R})|\delta - \delta'|^\alpha,
\end{aligned}$$

where the inequality follows from Lemma 1.4.1.

We take $\alpha \in \left(0, \frac{1}{2}\right)$ so that both estimates above hold. Plugging back to (1.4.2), we see that equality (1.4.1) holds.

We are now ready to verify that X_t has density u_t. In fact

$$\langle X_t, f \rangle = \lim_{\delta \to 0} \langle T_\delta X_t, f \rangle_0 = \langle u_t, f \rangle_0 .$$

□

We now derive the SPDE satisfied by the density field $u_t(x)$.

Theorem 1.4.3. *The density random field $u_t(x)$ of the SBM is a solution to the following SPDE:*

$$\partial_t u_t(x) = \frac{1}{2}\Delta u_t(x) + \sqrt{\gamma u_t(x)}\dot{W}_{tx}. \tag{1.4.3}$$

Namely, $u_t(x)$ is a weak solution of SPDE (1.4.3) in the following sense: for any $f \in C_0^2(\mathbb{R})$, we have

$$\langle u_t, f \rangle_0 = \langle u_0, f \rangle_0 + \int_0^t \left\langle u_s, \frac{1}{2} f'' \right\rangle_0 ds + \int_0^t \int_{\mathbb{R}} \sqrt{\gamma u_s(x)} f(x) W(dsdx), \tag{1.4.4}$$

where W is a white noise random measure on $\mathbb{R}_+ \times \mathbb{R}$ with intensity measure $dtdx$ (see Appendix A.1 for definition).

Proof. Let $\mathcal{S}'(\mathbb{R})$ be the space of Schwartz distributions. Then, M_t is an $\mathcal{S}'(\mathbb{R})$-valued continuous square-integrable martingale with

$$\langle M(f) \rangle_t = \gamma \int_0^t \int_{\mathbb{R}} f(x)^2 u_s(x) dx ds.$$

Let $\gamma : \mathbb{R}_+ \times \Omega \to L_{(2)}(H, H)$ be defined as

$$\gamma(s, \omega) f(x) = \sqrt{\gamma u_s(x)} f(x), \qquad \forall f \in H,$$

where $H = L^2(\mathbb{R})$ and $L_{(2)}(H, H)$ is the space consisting of all Hilbert-Schmidt operators on H. By Theorem A.1.7, on an extension of the original stochastic basis, there exists an H-cylindric Brownian motion (see Appendix A.1 for definition) B_t such that

$$M_t(f) = \int_0^t \langle \gamma(s, \omega) f, dB_s \rangle_H .$$

Let $\{h_j\}$ be a complete, orthonormal system (CONS) of the Hilbert space H and define random measure W on $\mathbb{R}_+ \times \mathbb{R}$ as

$$W([0, t] \times A) = \sum_{j=1}^{\infty} \langle 1_A, h_j \rangle B_t^{h_j} .$$

It is easy to show that W is a white noise random measure on $\mathbb{R}_+ \times \mathbb{R}$ with Lebesgue measure as its intensity. Furthermore,

$$M_t(f) = \sqrt{\gamma} \int_0^t \int_{\mathbb{R}} \sqrt{u_s(x)} f(x) W(dsdx).$$

Plugging back to (1.1.8) verifies that u_t satisfies SPDE (1.4.4). □

Finally, we will prove the joint continuity of the random field $u_t(x)$. Similar to Lemma 1.4.1, we obtain further estimates of the heat kernel.

Lemma 1.4.4. *Let $\alpha \in (0,1)$. For any $t > s$ and $y \in \mathbb{R}$, we have*

$$\int_s^t \int_{\mathbb{R}} p_{t-r}(x)^2 dxdr \le K|t-s|^{1/2}, \tag{1.4.5}$$

$$\int_0^s \int_{\mathbb{R}} |p_{t-r}(x) - p_{s-r}(x)|^2 dxdr \le K|t-s|^{1/2}, \tag{1.4.6}$$

and

$$\int_0^T \int_{\mathbb{R}} |p_t(x-y) - p_t(x)|^2 dxdt \le K|y|^\alpha. \tag{1.4.7}$$

Proof. The first estimate follows from the following simple calculation directly

$$\begin{aligned}\int_s^t \int_{\mathbb{R}} p_{t-r}(x)^2 dxdr &= \int_s^t p_{2(t-r)}(0) dr \\ &= (2\pi)^{-1/2}\sqrt{t-s}.\end{aligned}$$

The second estimate also can be estimated directly as

$$\begin{aligned}&\int_0^s \int_{\mathbb{R}} |p_{t-r}(x) - p_{s-r}(x)|^2 dxdr \\ &= \int_0^s \left(p_{2(t-r)}(0) + p_{2(s-r)} - 2p_{t+s-2r}(0)\right) dr \\ &= \pi^{-1/2}\left(\sqrt{t} - \sqrt{t-s} + \sqrt{s} - \sqrt{2}\sqrt{t+s} + \sqrt{2}\sqrt{t-s}\right) \\ &\le K|t-s|^{1/2}.\end{aligned}$$

To prove (1.4.7), we note that

$$\begin{aligned}|p_t(x-y) - p_t(x)| &= \left|\int_0^y p_t(x-z)\frac{x-z}{t} dz\right| \\ &\le 4t^{-1/2} \int_0^{|y|} p_{2t}(x-z) dz \\ &\le 4t^{-1/2}|y| \left(p_{2t}(x-y) + p_{4t}(x)\right).\end{aligned}$$

Thus,

$$\begin{aligned}
&\int_0^T \int_{\mathbb{R}} |p_t(x-y) - p_t(x)|^2 dxdt \\
&\le K_1 \int_0^T \int_{\mathbb{R}} t^{-\alpha/2}|y|^\alpha \left(p_{2t}(x-y) + p_{4t}(x)\right)^\alpha \left(p_t(x-y) + p_t(x)\right)^{2-\alpha} dx \\
&\le K_2 \int_0^T |y|^\alpha t^{-\alpha/2} t^{-1/2} dt \le K_3 |y|^\alpha,
\end{aligned}$$

for α such that $\frac{\alpha}{2} + \frac{1}{2} < 1$, namely, $\alpha < 1$. □

The following estimate on the moments of the density random field is needed in proving the joint continuity of $u_t(x)$ in (t, x).

Lemma 1.4.5. *If ν is finite and satisfies*

$$\sup_{t,x} \langle \nu, p_t(x - \cdot)\rangle < \infty, \tag{1.4.8}$$

then

$$\sup_{t>0,\ x\in\mathbb{R}} \mathbb{E}u_t(x)^n < \infty \tag{1.4.9}$$

for all $n \in \mathbb{N}$.

Proof. We use the moment duality to prove (1.4.9). Let $f_0^\epsilon = p_\epsilon^{\otimes n}$ and denote the moment dual process by f_t^ϵ. Then,

$$\mathbb{E}\left\langle X_t^{\otimes n}, f_0^\epsilon\right\rangle = \mathbb{E}\left(\left\langle \nu^{\otimes n_t}, f_t^\epsilon\right\rangle \exp\left(\frac{1}{2}\int_0^t n_s(n_s-1)ds\right)\right).$$

Taking limit $\epsilon \downarrow 0$ and using Fatou's lemma, we have

$$\begin{aligned}
\mathbb{E}u_t(x)^n &\le \liminf_{\epsilon\to 0} \mathbb{E}\left(\left\langle \nu^{\otimes n_t}, f_t^\epsilon\right\rangle \exp\left(\frac{1}{2}\int_0^t n_s(n_s-1)ds\right)\right) \\
&\le \exp\left(\frac{1}{2}n(n-1)t\right) \liminf_{\epsilon\to 0} \sum_{i=1}^n \mathbb{E}\left(\left\langle \nu^{\otimes n_t}, f_t^\epsilon\right\rangle 1_{\tau_{i-1}\le t<\tau_i}\right).
\end{aligned}$$

Now we estimate the sum above. We will consider the term with $i = 3$ only because the other terms can be dealt with similarly.

Denote the left hand side of (1.4.8) by K_1 and the bound of $\sqrt{t}p_t(x)$ by K_2. Denote

$$\tilde{y}^{k\ell} = (y_1, \cdots, y_{n-2}, \cdots, y_{n-2}, \cdots, y_{n-3}),$$

where y_{n-2} is at the kth and the ℓth positions. Then

$$\begin{aligned}
&\mathbb{E} f_t^\epsilon(x_1, \cdots, x_{n-2}) \\
&\le K\mathbb{E} \int_{\mathbb{R}^{n-2}} \prod_{i=1}^{n-2} p_{t-\tau_2}(x_i - y_i) S_2 f_{\tau_2-}^\epsilon(y) dy \\
&= K\mathbb{E} \int_{\mathbb{R}^{n-2}} \prod_{i=1}^{n-2} p_{t-\tau_2}(x_i - y_i) \sum_{1\le k<\ell\le n-2} \frac{2}{(n-2)(n-3)} f_{\tau_2-}^\epsilon(\tilde{y}^{k\ell}) dy.
\end{aligned}$$

Since

$$\begin{aligned}
&\mathbb{E} f_{\tau_2-}^\epsilon(\tilde{y}^{k\ell}) \\
&\le K\mathbb{E} \int_{\mathbb{R}^{n-1}} \prod_{j=1}^{n-1} p_{\tau_2-\tau_1}(\tilde{y}_j^{k\ell} - z_j) S_1 f_{\tau_1}^\epsilon(z) dz \\
&= K\mathbb{E} \int_{\mathbb{R}^{n-1}} \prod_{j=1}^{n-1} p_{\tau_2-\tau_1}(\tilde{y}_j^{k\ell} - z_j) \sum_{1\le k'<\ell'\le n-1} f_{\tau_1-}^\epsilon(\tilde{z}^{k'\ell'}) \frac{2}{(n-1)(n-2)} dz \\
&= K\mathbb{E} \int_{\mathbb{R}^{n-1}} \prod_{j=1}^{n-1} p_{\tau_2-\tau_1}(\tilde{y}_j^{k\ell} - z_j) p_{\tau_1+\epsilon}(z_1 - x) \cdots \\
&\qquad \times p_{\tau_1+\epsilon}(z_{n-2} - x) p_{\tau_1+\epsilon}(z_{n-1} - x)^2 dz,
\end{aligned}$$

we have

$$\begin{aligned}
&\mathbb{E}\left\langle \nu^{\otimes n-2}, f_t^\epsilon \right\rangle \\
&\le K^2 \mathbb{E} \int_{\mathbb{R}^{n-2}} \prod_{i=1}^{n-2} \int_{\mathbb{R}} p_{t-\tau_2}(x_i - y_i) \nu(dx_i) \sum_{1\le k<\ell\le n-2} \frac{2}{(n-2)(n-3)} \\
&\quad \times \int_{\mathbb{R}^{n-1}} \prod_{j=1}^{n-1} p_{\tau_2-\tau_1}(\tilde{y}_j^{k\ell} - z_j) p_{\tau_1+\epsilon}(z_1 - x) \cdots \\
&\qquad \times p_{\tau_1+\epsilon}(z_{n-2} - x) p_{\tau_1+\epsilon}(z_{n-1} - x)^2 dz dy.
\end{aligned}$$

Using (1.4.8), we continue the estimate as

$$\begin{aligned}
&\mathbb{E}\left\langle \nu^{\otimes n-2}, f_t^\epsilon \right\rangle \\
&\le K^2 K_1^{n-3} \mathbb{E} \int_{\mathbb{R}^{n-2}} \int_{\mathbb{R}} p_{t-\tau_2}(x_{n-2} - y_{n-2}) \nu(dx_{n-2}) \\
&\quad \times \sum_{1\le k<\ell\le n-2} \frac{2}{(n-2)(n-3)} \int_{\mathbb{R}^{n-1}} \prod_{j=1}^{n-1} p_{\tau_2-\tau_1}(\tilde{y}_j^{k\ell} - z_j) \\
&\quad \times p_{\tau_1+\epsilon}(z_1 - x) \cdots p_{\tau_1+\epsilon}(z_{n-2} - x) p_{\tau_1+\epsilon}(z_{n-1} - x)^2 dz dy.
\end{aligned}$$

By the definition of K_2, we further estimate $\mathbb{E}\left\langle \nu^{\otimes n-2}, f_t^\epsilon \right\rangle$ by

$$
\begin{aligned}
&K^2K_1^{n-3}\mathbb{E}\int_{\mathbb{R}}\int_{\mathbb{R}} p_{t-\tau_2}(x_{n-2}-y_{n-2})\nu(dx_{n-2}) \sum_{1\le k<\ell\le n-2} \frac{2}{(n-2)(n-3)} \\
&\quad\times \int_{\mathbb{R}^{n-1}} \frac{K_2}{\sqrt{\tau_2-\tau_1}} p_{\tau_2-\tau_1}(y_{n-2}-z_k)p_{\tau_1+\epsilon}(z_1-x)\cdots \\
&\qquad\qquad \times p_{\tau_1+\epsilon}(z_{n-2}-x)p_{\tau_1+\epsilon}(z_{n-1}-x)\frac{K_2}{\sqrt{\tau_1+\epsilon}}dzdy_{n-2} \\
&\le \mathbb{E}\frac{K^2K_1^{n-3}K_2^2}{\sqrt{\tau_1(\tau_2-\tau_1)}}\int_{\mathbb{R}}\int_{\mathbb{R}} p_{t-\tau_2}(x_{n-2}-y_{n-2})\nu(dx_{n-2}) \\
&\quad\times \sum_{1\le k<\ell\le n-2} \frac{2}{(n-2)(n-3)} p_{\tau_2+\epsilon}(y_{n-2}-x)dy_{n-2} \\
&\le \mathbb{E}\frac{K^2K_1^{n-3}K_2^2}{\sqrt{\tau_1(\tau_2-\tau_1)}}\int_{\mathbb{R}} p_{t+\epsilon}(x_{n-2}-x)\nu(dx_{n-2}).
\end{aligned}
$$

Therefore

$$
\mathbb{E}\left\langle \nu^{\otimes n_t}, f_t^\epsilon\right\rangle 1_{\tau_2\le t<\tau_3} \le K^2K_1^{n-3}K_2^2\nu(\mathbb{R})\mathbb{E}\frac{1}{\sqrt{\tau_1(\tau_2-\tau_1)}} < \infty.
$$

As we mentioned earlier, the other terms can be proved similarly. □

Finally, we are ready to state and prove the main result of this section.

Theorem 1.4.6. *The density random field $u_t(x)$ is jointly continuous in (t,x) almost surely.*

Proof. Using the convolution form, $u_t(x)$ can be represented as

$$
u_t(x) = \int_{\mathbb{R}} p_t(y-x)\nu(dy) + \int_0^t\int_{\mathbb{R}} p_{t-s}(y-x)\sqrt{\gamma u_s(y)}W(dsdy).
$$

Namely, $u_t(x)$ is a mild solution of SPDE (1.4.3). The joint continuity of the first term is easy, so we focus on the second, which we denote by $v_t(x)$.

For $s<t$ and positive integer n, we have

$$
\begin{aligned}
&\mathbb{E}|v_t(x)-v_s(x)|^{2n} \\
&\le K_1\mathbb{E}\left|\int_0^s\int_{\mathbb{R}} \left(p_{t-r}(y-x)-p_{s-r}(y-x)\right)\sqrt{u_r(y)}W(drdy)\right|^{2n} \\
&\quad + K_1\mathbb{E}\left|\int_s^t\int_{\mathbb{R}} p_{t-r}(y-x)\sqrt{u_r(y)}W(drdy)\right|^{2n}.
\end{aligned}
$$

By Burkholder's inequality, we continue the estimate as

$$\begin{aligned}
&\mathbb{E}|v_t(x) - v_s(x)|^{2n} \\
&\le K_2\mathbb{E}\left|\int_0^s \int_{\mathbb{R}} (p_{t-r}(y-x) - p_{s-r}(y-x))^2 u_r(y)dydr\right|^n \\
&+K_2\mathbb{E}\left|\int_s^t \int_{\mathbb{R}} p_{t-r}(y-x)^2 u_r(y)dydr\right|^n .
\end{aligned}$$

Making use of Lemma 1.4.5, we further estimate as

$$\begin{aligned}
&\mathbb{E}|v_t(x) - v_s(x)|^{2n} \\
&\le K_3\left|\int_0^s \int_{\mathbb{R}} (p_{t-r}(y-x) - p_{s-r}(y-x))^2 dydr\right|^n \\
&+K_3\left|\int_s^t \int_{\mathbb{R}} p_{t-r}(y-x)^2 dydr\right|^n .
\end{aligned}$$

Finally, by Lemma 1.4.4, we get

$$\mathbb{E}|v_t(x) - v_s(x)|^{2n} \le K_4|t-s|^{n/2}. \tag{1.4.10}$$

Similarly, we can prove that

$$\mathbb{E}|v_t(x) - v_t(y)|^{2n} \le K_5|x-y|^{n\alpha}. \tag{1.4.11}$$

Taking n large enough such that $n\alpha > 2$ (and hence $n > 4$), the conclusion of the theorem follows from (1.4.10), (1.4.11), and from Kolmogorov's criteria. □

1.5 The SPDE for the distribution

Uniqueness in distribution for the solution of SPDE (1.4.3) follows from that of the corresponding martingale problem. However, the existence and uniqueness of the strong solution remains a challenging open problem. To overcome this difficulty, we shall consider another SPDE which will be satisfied by the distribution function-valued process of the super-Brownian motion, for which the strong uniqueness of the solution can be derived.

We consider the "distribution function" valued process

$$u_t(x) \equiv X_t((-\infty, x]), \qquad \forall\, x \in \mathbb{R}. \tag{1.5.1}$$

It follows from integration by parts that

$$\langle X_t, f\rangle = -\langle u_t, f'\rangle_0 .$$

Then, it is easy to show that

$$\left\langle X_t, \frac{1}{2}f''\right\rangle = -\left\langle u_t, \frac{1}{2}f'''\right\rangle_0.$$

Furthermore, by substitution, we get

$$\begin{aligned}\langle X_t, f^2\rangle &= \int_0^t \int_{\mathbb{R}} f^2(x)du_s(x)ds \\ &= \int_0^t \int_0^\infty f^2(u_s^{-1}(b))dbds \\ &= \int_0^t \int_0^\infty \left(\int_{\mathbb{R}} f'(x)1_{b\le u_s(x)}dx\right)^2 dbds.\end{aligned}$$

By martingale representation theorem, there exists a white noise random measure B on $\mathbb{R}_+ \times \mathbb{R}_+$ with Lebesgue measure as the intensity such that

$$M_t(f) = \sqrt{\gamma}\int_0^t \int_0^\infty \int_{\mathbb{R}} 1_{b\le u_s(x)}f'(x)dxB(dsdb).$$

Summarize the above arguments, we find that $u_t(x)$ satisfies the following SPDE:

$$u_t(x) = \nu(x) + \int_0^t \frac{1}{2}\Delta u_s(x)ds + \sqrt{\gamma}\int_0^t \int_0^\infty 1_{b\le u_s(x)}B(dsdb). \qquad (1.5.2)$$

On the other hand, if $u_t(x)$ is a solution to SPDE (1.5.2), it is easy to show that the corresponding measure-valued process determined by (1.5.1) is a solution to martingale problem (1.1.8, 1.1.9).

Thus, we have proved the following theorem.

Theorem 1.5.1. *The measure-valued process X_t is a solution to martingale problem (1.1.8, 1.1.9) if and only if the random field $\{u_t(x)\}$ defined by (1.5.1) is a solution to SPDE (1.5.2).*

For the rest of the book, we shall study three classes of nonlinear SPDEs which are generalizations of PDE (1.2.2), SPDEs (1.4.3) and (1.5.2).

1.6 Historical remarks

As we pointed out at the beginning of this chapter, superprocesses were first studied by Jirina (1958), (1964) and Watanabe (1968). The systematical investigation started by Dawson (1975). There are many nice monographs and survey papers in this field available in literature. Here we mention a few: Dawson (1993), Dynkin (1994), Dynkin (2002), Etheridge (2000), LeGall (1999), Li (2011), Perkins (2002).

The main topic we presented here is the SPDE satisfied by the density random field of the superprocess. This part is based on the papers of Konno and Shiga (1988) and Reimers (1989) where this SPDE was derived and studied. The uniqueness of the solution to SPDE (1.4.3) is only proved in the weak sense using that of the MP. Many attempts have been made towards proving the strong uniqueness for the solution to (1.4.3). The main difficulty is the non-Lipschitz coefficient in front of the noise. Some progress has been made by relaxing the form of the SPDE. When the space $\mathbb{R}$ is replaced by a single point, (1.4.3) becomes a stochastic differential equation (SDE) which is the Feller's diffusion $dv_t = \sqrt{v_t}dB_t$ whose uniqueness is established using the Yamada-Watanabe argument. When the random field B is colored in space and white in time, the strong uniqueness of the solution to SPDE (1.4.3) with $\sqrt{v_t(x)}$ replaced by a function of $v_t(x)$ was obtained by Mytnik, Perkins and Sturm (2006) under suitable conditions. When B is a space-time white noise, Mytnik and Perkins (2011) proved pathwise uniqueness for multiplicative noises of the form $\sigma(x, v_t(x))\dot{B}_{tx}$, where σ is Hölder continuous of index $\alpha > \frac{3}{4}$ in the solution variable. In particular, their results imply that the following SPDE

$$\partial_t v_t(x) = \frac{1}{2}\Delta v_t(x) + |v_t(x)|^\alpha \dot{B}_{tx}, \tag{1.6.1}$$

has a pathwise unique solution when $\alpha > \frac{3}{4}$. Some negative results have also been achieved. When signed solutions are allowed, Mueller, Mytnik and Perkins (2011) gave a non-uniqueness result for SPDE (1.6.1) when $\frac{1}{2} \le \alpha < \frac{3}{4}$. When SPDE (1.6.1) is restricted to non-negative solutions, Burdzy, Mueller and Perkins (2010) showed a non-uniqueness result in the case of $0 < \alpha < \frac{1}{2}$.

Chapter 2

Superprocesses in Random Environments

In this chapter, we generalize the branching particle systems introduced in Chapter 1. Namely, we assume that the motions of the particles are governed not only by the individual Brownian motions $\{B_\alpha(t),\ \alpha \in \mathcal{I}\}$ but also by a common white noise random measure W defined on $\mathbb{R}_+ \times U$, where $(U, \mathcal{U}, \mu)$ is a σ-finite measure space representing the random factors affecting the whole system. The measure space $(U, \mathcal{U}, \mu)$ being σ-finite means that there exists a sequence U_n increasing to U such that $\mu(U_n) < \infty$ for all n.

2.1 Introduction and main result

First, we describe the particle system. We assume that the motion of the particle $\alpha \in \mathcal{I}$ (during her lifetime) is governed by the following SDE:

$$d\xi_\alpha(t) = dB_\alpha(t) + \int_U h(y, \xi_\alpha(t)) W(dtdy), \tag{2.1.1}$$

where $h:\ U \times \mathbb{R} \to \mathbb{R}$ is a measurable mapping and W is a white noise random measure on $\mathbb{R}_+ \times U$ with intensity measure μ (see Appendix A.1 for definition).

To obtain a unique strong solution to (2.1.1), we impose the following Lipschitz and linear growth conditions on h. That is, there exists a constant K such that $\forall\ x,\ y \in \mathbb{R}$,

$$\int_U |h(u,x) - h(u,y)|^2 \mu(du) \le K|x-y|^2, \tag{2.1.2}$$

and

$$\int_U |h(u,x)|^2 \mu(du) \le K(1+|x|^2). \tag{2.1.3}$$

Under these conditions, similar to the classical theory of stochastic differential equations we can prove that SDE (2.1.1) has a unique solution ξ_α for each $\alpha \in \mathcal{I}$.

The empirical measure process of the system as in (1.1.1) is given by

$$X_t^n = \frac{1}{n} \sum_{\alpha \sim_n t} \delta_{\xi_\alpha(t)}.$$

To prove the tightness of the sequence $\{X^n\}$, we follow the same procedure as in Section 1.1. For any $f \in C_b^2(\mathbb{R})$, Itô's formula is applied to $f(\xi_\alpha(t))$ to obtain

$$\begin{aligned} df(\xi_\alpha(t)) &= Lf(\xi_\alpha(t))dt + f'(\xi_\alpha(t))dB_\alpha(t) \\ &\quad + \int_U f'(\xi_\alpha(t))h(y, \xi_\alpha(t))W(dtdy), \end{aligned}$$

where

$$Lf = \frac{1}{2} a f'' \text{ and } a = 1 + \int_U h(y,x)^2 \mu(dy).$$

Summing up over all α alive at time t, we get

$$\begin{aligned} d\langle X_t^n, f\rangle &= \langle X_t^n, Lf\rangle\, dt + \frac{1}{n} \sum_{\alpha \sim_n t} f'(\xi_\alpha(t))dB_\alpha(t) \\ &\quad + \int_U \langle X_t^n, f'h(y,\cdot)\rangle\, W(dtdy). \end{aligned} \tag{2.1.4}$$

At a jumping time t, we have

$$\langle X_t^n - X_{t-}^n, f\rangle = \frac{1}{n} \sum_{\alpha_d = t} (\alpha_\# - 1) f(\xi_\alpha(t)). \tag{2.1.5}$$

Combining (2.1.4) and (2.1.5), we obtain

$$\langle X_t^n, f\rangle = \langle X_0^n, f\rangle + Y_t^n(f) + R_t^n(f) + U_t^n(f) + V_t^n(f),$$

where

$$Y_t^n(f) = \int_0^t \langle X_s^n, Lf\rangle\, ds,$$

$$R_t^n(f) = \int_0^t \int_U \langle X_s^n, f'h(y,\cdot)\rangle\, W(dsdy),$$

$$U_t^n(f) = \frac{1}{n} \int_0^t \sum_{\alpha \sim_n s} f'(\xi_\alpha(s))dB_\alpha(s),$$

and

$$V_t^n(f) = \frac{1}{n}\sum_{s\leq t}\sum_{\alpha_d=s}(\alpha_{\#}-1)f(\xi_\alpha(s)).$$

Note that $R_t^n(f)$, $U_t^n(f)$ and $V_t^n(f)$ are three uncorrelated martingales with quadratic variation processes

$$\langle R^n(f)\rangle_t = \int_0^t \int_U |\langle X_s^n, f'h(y,\cdot)\rangle|^2 dyds,$$

$$\langle U^n(f)\rangle_t = \frac{1}{n}\int_0^t \left\langle X_s^n, |f'|^2\right\rangle ds$$

and

$$\langle V^n(f)\rangle_t = \int_0^t \langle X_r^n, \gamma f^2\rangle\, dr.$$

Proofs for the tightness of Y^n, U^n, V^n are the same as those in Chapter 1, while the tightness of R^n follows from similar arguments. Further, it is easy to show that any limit point X is a solution to the martingale problem (2.1.6, 2.1.7) given below.

The rest of this chapter will be devoted to the proof of the following convergence result.

Theorem 2.1.1. *Assume that $\nu^n \to \nu$ in $\mathcal{M}_F(\mathbb{R})$ and*

$$\sup_n \int_{\mathbb{R}} \left(1+x^2\right)\nu^n(dx) < \infty.$$

Then, sequence $\{X^n\}$ converges weakly in $D([0,\infty),\mathcal{M}_F(\mathbb{R}))$. Furthermore, the limit process X lies in $C(\mathbb{R}_+,\mathcal{M}_F(\mathbb{R}))$ and is the unique solution of the following martingale problem (MP): For all $f \in C_b^2(\mathbb{R})$,

$$M_t(f) \equiv \langle X_t, f\rangle - \langle \nu, f\rangle - \int_0^t \langle X_s, Lf\rangle\, ds \tag{2.1.6}$$

is a continuous square integrable martingale with initial $M_0(f) = 0$ and the quadratic variation process

$$\langle M(f)\rangle_t = \int_0^t \gamma\left\langle X_s, f^2\right\rangle ds + \int_0^t \int_U \left|\langle X_s, h(y,\cdot)f'\rangle\right|^2 \mu(dy)ds. \tag{2.1.7}$$

2.2 The moment duality

To prove the uniqueness of the solution to martingale problem (2.1.6, 2.1.7), we establish a moment duality relation between the solution of the martingale problem and a function-valued process. The construction is similar to that of Section 1.3. The only difference is that the n-particle movements are no longer independent in the current setup.

Let $\{X_t\}$ be a solution to martingale problem (2.1.6, 2.1.7).

Lemma 2.2.1. *For any $f \in C_b^2(\mathbb{R}^n)$, the stochastic process*

$$F(X_t,(n,f)) - \int_0^t G(X_s,(n,f))ds \tag{2.2.1}$$

is a martingale, where for any every $\mu \in \mathcal{M}_F(\mathbb{R})$,

$$F(\mu,(n,f)) = \langle \mu^{\otimes n}, f\rangle \equiv \int_{\mathbb{R}} \cdots \int_{\mathbb{R}} f(x_1,\cdots,x_n)\mu(dx_1)\cdots\mu(dx_n)$$

and

$$G(\mu,(n,f)) = \left\langle \mu^{\otimes n}, \mathcal{A}^{(n)} f\right\rangle + \gamma \sum_{1\le j<k\le n} \left\langle \mu^{\otimes(n-1)}, \Phi_{jk} f\right\rangle,$$

while $\mathcal{A}^{(n)}$ is the generator of the Markov process describing the movement of n particles in the system, i.e.,

$$\begin{aligned}\mathcal{A}^{(n)} f(x_1,\cdots,x_n) = &\sum_{1\le j<k\le n} \rho(x_j,x_k)\partial_j\partial_k f(x_1,\cdots,x_n)\\ &+\sum_{i=1}^n L_i f(x_1,\cdots,x_n),\end{aligned}$$

with

$$\rho(x,z) = \int_U h(y,x)h(y,z)\mu(dy),$$

$L_j f$ means that the operator L is applied to the jth variable while keeping the other variables constant, and $\Phi_{jk} f$ is the same as that defined in Section 1.3.

Proof. Again, using approximation if necessary, we may and will assume that

$$f(x_1,\cdots,x_n) = \prod_{j=1}^n f_j(x_j), \qquad f_j \in C_b^2(\mathbb{R}).$$

Similar to Section 1.3, it is easy to show that

$$\begin{aligned}\langle M(f), M(g)\rangle_t = &\int_0^t \int_U \langle X_s, h(y,\cdot)f'\rangle \langle X_s, h(y,\cdot)g'\rangle)\mu(dy)ds \\ &+ \gamma \int_0^t \langle X_s, fg\rangle\, ds.\end{aligned}$$

Applying Itô's formula, we have

$$\begin{aligned}&dF(X_t,(n,f)) \\ &= \sum_{j=1}^n \left(\prod_{i\neq j} \langle X_t, f_i\rangle\right) \left\langle X_t, \frac{1}{2}f_j''\right\rangle dt \\ &\quad + \gamma \sum_{j<k} \left(\prod_{i\notin\{j,k\}} \langle X_t, f_i\rangle\right) \langle X_t, f_j f_k\rangle\, dt \\ &\quad + \sum_{j<k} \left(\prod_{i\notin\{j,k\}} \langle X_t, f_i\rangle\right) \int_U \langle X_t, h(y,\cdot)f_j'\rangle \langle X_t, h(y,\cdot)f_k'\rangle\, \mu(dy)dt \\ &\quad + d(mart.) \\ &= G(X_t,(n,f))dt + d(mart.),\end{aligned}$$

which proves that (2.2.1) is a martingale. □

Let E_1 and E_2 be the same as those given in Section 1.3. Again, let n_t be Kingman's coalescent process and S_k denote the same process as that defined in Section 1.3. The only thing needed to be changed is the semigroup. Denote by $T_t^{(n)}$ the semigroup generated by $\mathcal{A}^{(n)}$, i.e., the semigroup corresponding to the n-particle movement Markov process. Let $\tau_0 = 0$, $\tau_{n_0} = \infty$ and $\{\tau_k : 1 \le k < n_0\}$ be the sequence of jumping times of $\{n_t : t \ge 0\}$. For $\tau_k \le t < \tau_{k+1}$, define

$$f_t = T_{t-\tau_k}^{(n_{\tau_k})} S_k T_{\tau_k-\tau_{k-1}}^{(n_{\tau_{k-1}})} S_{k-1} \cdots T_{\tau_2-\tau_1}^{(n_{\tau_1})} S_1 T_{\tau_1}^{(n_0)} f_0.$$

Then, for any $\mu \in \mathcal{M}_F(\mathbb{R})$,

$$F(\mu,(n_t,f_t)) - \int_0^t H(\mu,(n_s,f_s))ds \tag{2.2.2}$$

is a martingale, where

$$H(\mu,(n,f)) = G(\mu,(n,f)) + \frac{1}{2}\gamma n(n-1)F(\mu,(n,f)).$$

Combining (2.2.1) and (2.2.2), the following formula follows from Theorem 1.3.1 directly.

Theorem 2.2.2. *For any $t \geq 0$, $n \in \mathbb{N}$ and $f \in C_b(\mathbb{R}^n)$, we have*

$$\mathbb{E}F(X_t, (n, f)) = \mathbb{E}\left(F(\mu, (n_t, f_t)) \exp\left(\int_0^t \frac{\gamma}{2} n_s(n_s - 1) ds\right)\right).$$

Corollary 2.2.3. *The martingale problem (2.1.6, 2.1.7) is well-posed.*

Proof. Similar to the proof of Corollary 1.3.6, we have

$$\alpha_n \equiv \mathbb{E}\left(\langle X_t, \phi\rangle^n\right) \leq \tilde{\alpha}_n \|\phi\|_\infty^n,$$

where $\tilde{\alpha}_n = \mathbb{E}\left(\langle X_t, 1\rangle^n\right)$. The rest of the proof follows from exactly the same arguments as those in the proof of Corollary 1.3.6. □

2.3 Conditional martingale problem

In the study of stochastic processes in random environment, it is usually convenient to fix the environment first. In such an approach, the properties obtained are usually called the quenched properties of the processes. In this section, we study the martingale problem with the environment W fixed. Such a martingale problem is called the quenched martingale problem or the conditional martingale problem (CMP).

Definition 2.3.1. A real valued process U_t (adapted to σ-field $\mathcal{F}_t$) is a P^W-martingale if for any $t > s$,

$$\mathbb{E}\left(U_t | \mathcal{F}_s \vee \sigma(W)\right) = U_s, \qquad \text{a.s.}$$

Theorem 2.3.2. *X_t is a solution to MP (2.1.6, 2.1.7) if and only if it is a solution to the following conditional martingale problem: For any $f \in C_0^2(\mathbb{R})$,*

$$\begin{aligned} N_t(f) \equiv \langle X_t, f\rangle - \langle \nu, f\rangle - \int_0^t \langle X_s, Lf\rangle\, ds \\ - \int_0^t \int_U \langle X_s, h(y, \cdot) f'\rangle\, W(dsdy) \end{aligned} \tag{2.3.1}$$

is a continuous P^W-martingale with quadratic variation process

$$\langle N(f)\rangle_t = \gamma \int_0^t \langle X_s, f^2\rangle\, ds. \tag{2.3.2}$$

Proof. Recall that $\mathcal{S}'(\mathbb{R})$ is the space of Schwartz distributions. Suppose that X_t is a solution to MP (2.1.6, 2.1.7). Then, M_t is a $\mathcal{S}'(\mathbb{R})$-valued martingale with quadratic variation process

$$\langle M(f)\rangle_t = \int_0^t \left(\|\langle X_s, h(\bullet, \cdot)f'\rangle\|^2_{L^2(U,\mu)} + \|\phi(s, X_s)f\|_0^2 \right) ds,$$

where $\phi(s, X_s)$ is a linear map from $L^2(\mathbb{R})$ to $\mathcal{S}'(\mathbb{R})$ such that

$$\langle X_s, f_1 f_2\rangle = \langle \phi(s, X_s)f_1, \phi(s, X_s)f_2\rangle_0, \qquad \forall\ f_1.\ f_2 \in \mathcal{S}(\mathbb{R}).$$

By the martingale representation Theorem A.1.8 there exist independent processes $\tilde{W}$ and B such that $\tilde{W}$ is an $L^2(U, \mu)$-c.B.m., B is an H_0-cylindrical Brownian motion (see Appendix A.1 for definition), and

$$M_t(f) = \int_0^t \left\langle \langle X_s, h(\bullet, \cdot)f'\rangle, d\tilde{W}_s \right\rangle_{L^2(U,\mu)} + \int_0^t \langle \phi(s, X_s)f, dB_s\rangle_0,$$

where $\phi(s, X_s)$ is a linear mapping from H_0 to $\mathcal{S}'(\mathbb{R})$ such that

$$\langle X_s, f_1 f_2\rangle = \langle \phi(s, X_s)f_1, \phi(s, X_s)f_2\rangle_0, \qquad \forall\ f_1.\ f_2 \in \mathcal{S}(\mathbb{R}).$$

Define random measure W on $\mathbb{R}_+ \times U$ as

$$W([0, t] \times A) = \sum_{j=1}^{\infty} \tilde{W}_t(\phi^j) \left\langle 1_A, \phi^j \right\rangle_{L^2(U,\mu)}, \quad \forall\ A \in \mathcal{U}.$$

Then, W is a white noise random measure with intensity μ such that

$$\int_0^t \left\langle \langle X_s, h(\bullet, \cdot)f'\rangle, d\tilde{W}_s \right\rangle_{L^2(U,\mu)} = \int_0^t \int_U \langle X_s, h(y, \cdot)f'\rangle\, W(dsdy).$$

Hence, X_t solves the CMP.

On the other hand, suppose that X_t is a solution to the CMP. As $N_t(f)$ is a P^W-martingale, for $s < t$, we have

$$\begin{aligned}\mathbb{E}(N_t(f)\tilde{W}_t(g)|\mathcal{F}_s^X) &= \mathbb{E}\left(\mathbb{E}\left(N_t(f)|\mathcal{F}_s \vee \sigma(W)\right) \tilde{W}_t(g)|\mathcal{F}_s^X\right) \\ &= \mathbb{E}\left(N_s(f)\tilde{W}_t(g)|\mathcal{F}_s^X\right) \\ &= N_s(f)\tilde{W}_s(g),\end{aligned}$$

where $\mathcal{F}_t^X$ is the σ-field generated by X. Hence, the quadratic covariation process $\left\langle N(f), \tilde{W}(g)\right\rangle_0 = 0$. Therefore,

$$M_t(f) = N_t(f) + \int_0^t \left\langle \langle X_s, h(\bullet, \cdot)f'\rangle, d\tilde{W}_s \right\rangle_{L^2(U,\mu)}$$

is a martingale with quadratic variation process

$$\begin{aligned}\langle M(f)\rangle_t &= \langle N(f)\rangle_t + \int_0^t \left\langle \langle X_s, h(\bullet,\cdot)f'\rangle, d\tilde{W}_s\right\rangle^2_{L^2(U,\mu)} \\ &= \int_0^t \gamma X_s(f^2)ds + \int_0^t \int_U |\langle X_s, h(y,\cdot)f'\rangle|^2 \mu(dy)ds.\end{aligned}$$

This proves that X_t is a solution to the MP (2.1.6, 2.1.7). □

Finally, as a warm up, we derive an SPDE arising from the superprocess in random environment. This is one of the three classes of nonlinear SPDEs arising from SPRE.

We consider the "distribution function" valued process

$$u_t(x) \equiv X_t((-\infty, x]), \qquad \forall\, x \in \mathbb{R}. \tag{2.3.3}$$

We will use the following convention on the notation ∇: When f is a function on $\mathbb{R}\times H$ where H is a space other than $\mathbb{R}$, $\nabla f(x,y)$ denotes the partial derivative with respect to spatial variable x, namely, $\nabla f(x,y) = \partial_x f(x,y)$ for any $x\in\mathbb{R}$ and $y\in H$. When $H=\mathbb{R}$, we will use notation ∇_x or ∇_y to indicate the variable with respect to which we are taking the partial derivative. When f is a function on $\mathbb{R}$, we will simply use f' to denote the derivative.

As in Section 1.5, we note that

$$\langle X_t, f\rangle = -\langle u_t, f'\rangle_0, \qquad \forall\, f\in C_0^1(\mathbb{R}).$$

Then, it is easy to show that $\forall\, f\in C_0^3(\mathbb{R})$,

$$\langle X_t, Lf\rangle = -\langle u_t, L_2 f'\rangle_0 \text{ and } \langle X_t, h(y,\cdot)f\rangle = -\langle u_t, L_{y,1}f'\rangle,$$

where

$$L_{y,1}g = h(y,\cdot)g' + \nabla h(y,\cdot)g \text{ and } L_2 g = \frac{1}{2}ag'' + a'g'.$$

Furthermore, by substitution, we get

$$\begin{aligned}\langle X_t, f^2\rangle &= \int_{\mathbb{R}} f^2(x)du_t(x) \\ &= \int_0^\infty f^2(u_t^{-1}(y))dy \\ &= \int_0^\infty \left(\int_{\mathbb{R}} f'(x)1_{y\le u_t(x)}dx\right)^2 dy.\end{aligned}$$

Similar to the proof of Theorem 2.3.2, there exists a white noise random measure B on $\mathbb{R}_+ \times \mathbb{R}_+$ with the Lebesgue measure as the intensity such that

$$N_t(f) = \gamma \int_0^t \int_0^\infty \int_{\mathbb{R}} 1_{y \le u_s(x)} f'(x) dx B(dsdy).$$

Summarize the above statements, we find that $u_t(x)$ satisfies the following SPDE:

$$\begin{aligned} u_t(x) = \nu(x) + \int_0^t L_2^* u_s(x) ds + \int_0^t \int_U L_{y,1}^* u_s(x) W(dsdy) \\ + \gamma \int_0^t \int_0^\infty 1_{y \le u_s(x)} B(dsdy). \end{aligned} \tag{2.3.4}$$

On the other hand, if $u_t(x)$ is a solution to SPDE (2.3.4), it is not difficult to show that the corresponding measure-valued process determined by (2.3.3) is a solution to martingale problem (2.1.6, 2.1.7).

Thus, we have proved the following theorem.

Theorem 2.3.3. *The measure-valued process X_t is a solution to martingale problem (2.1.6, 2.1.7) if and only if the random field $u_t(x)$ defined by (2.3.3) is a solution to SPDE (2.3.4).*

2.4 Historical remarks

We would like to point out the relationship between our model and those studied in the literature. If $h = 0$, X_t is the classical SBM which was introduced in the previous chapter. On the other hand, if $U = \mathbb{R}$ and $h(y, x) = \tilde{h}(y - x)$, X_t is investigated by Wang (1998) and Dawson, Li, Wang (2001); if U consists of finitely many points only, then W is a finite dimensional Brownian motion and X_t is the process studied by Skoulakis and Adler (2001).

We also note that when $\gamma = 0$, X_t is related to the unnormalized filter in the theory of stochastic filtering. An SPDE corresponding to the MP (2.1.6, 2.1.7) is called the Duncan-Mortensen-Zakai equation which is established by Duncan (1967), Mortensen (1966) and Zakai (1969), and has been studied by many others (cf. Kallianpur (1980)). The normalized version of this SPDE is a nonlinear SPDE satisfied by the optimal filter. This nonlinear SPDE is called the Kushner-FKK equation in attribution to the work of Kushner (1967) and Fujisaki *et al* (1972). We refer the reader to

Bain and Crisan (2009) and Xiong (2008) for an introduction and for the recent development of this topic.

Chapter 3

Linear SPDE

As a preparation for the study of the three classes of nonlinear SPDEs related to the superprocesses in random environments, in this chapter, we consider a superprocess in a random environment without branching. This process satisfies the following linear SPDE: $\forall\ \phi \in C_b^2(\mathbb{R})$,

$$\langle X_t, \phi\rangle = \langle \nu, \phi\rangle + \int_0^t \langle X_s, L\phi\rangle\, ds + \int_0^t \int_U \langle X_s, h(y,\cdot)\phi'\rangle\, W(dsdy). \quad (3.0.1)$$

We will study some properties for the linear SPDE (3.0.1) and establish a duality relationship between this SPDE and one whose solution is represented by the conditional (given the random environment) transition probabilities of a typical particle in the population introduced in the last chapter.

3.1 An equation on measure space

In this section, we consider a generalized version of SPDE (3.0.1). Let $b,\ d,\ \sigma : \mathbb{R}_+ \times \mathbb{R} \times \Omega \to \mathbb{R}$ and $\alpha,\ \beta : \mathbb{R}_+ \times \mathbb{R} \times \Omega \to L^2(U,\mu)$ be random mappings satisfying the following uniform boundedness condition.

Assumption (UB): There exists a constant K such that for any $(t,x,\omega) \in \mathbb{R}_+ \times \mathbb{R} \times \Omega$, we have

$$|b_t(x)|^2 + |d_t(x)|^2 + |\sigma_t(x)|^2 + \int_U \left(\alpha_t(x,y)^2 + \beta_t(x,y)^2\right) \mu(dy) \le K.$$

Let $\mathcal{M}_G(\mathbb{R})$ be the space of finite signed measures on $\mathbb{R}$. We consider the following linear equation on $\mathcal{M}_G(\mathbb{R})$:

$$\begin{aligned}\langle U_t, \phi\rangle = \langle U_0, \phi\rangle &+ \int_0^t \langle U_s, d_s\phi + L_s\phi\rangle\, ds \qquad (3.1.1)\\ &+ \int_0^t \int_U \langle U_s, \beta_s(\cdot,y)\phi + \alpha_s(\cdot,y)\phi'\rangle\, W(dyds),\end{aligned}$$

where for $s \geq 0$ and $x \in \mathbb{R}$,

$$L_s\phi(x) = \frac{1}{2}a_s(x)\phi''(x) + b_s(x)\phi'(x),$$

and

$$a_s(x) = a_s^{(1)}(x) + a_s^{(2)}(x) = \sigma_s(x)^2 + \int_U \alpha_s(x,y)^2\mu(dy).$$

Recall that $H_0 = L^2(\mathbb{R})$ is the Hilbert space with the usual L^2-norm $\|\cdot\|_0$ and inner product $\langle\cdot,\cdot\rangle_0$ given by

$$\|\phi\|_0^2 = \int_{\mathbb{R}} |\phi(x)|^2 dx$$

and

$$\langle\phi,\psi\rangle_0 = \int_{\mathbb{R}} \phi(x)\psi(x)dx.$$

To obtain good estimates and derive uniqueness for the solution to (3.1.1), we transform an $\mathcal{M}_G(\mathbb{R})$-valued process to an H_0-valued process. For any $\nu \in \mathcal{M}_G(\mathbb{R})$ and $\delta > 0$, let

$$(T_\delta\nu)(x) = \int_{\mathbb{R}} p_\delta(x-y)\nu(dy). \tag{3.1.2}$$

We also use the same notation for the Brownian semigroup on H_0, i.e.,

$$T_t\phi(x) = \int_{\mathbb{R}} p_t(x-y)\phi(y)dy, \qquad \forall\phi \in H_0.$$

We will need the following simple facts.

Lemma 3.1.1. *i) If $\phi \in H_0$ and $\delta > 0$, then $\|T_\delta\phi\|_0 \leq \|\phi\|_0$.*
ii) If $\nu \in \mathcal{M}_G(\mathbb{R})$ and $\delta > 0$, then $T_\delta\nu \in H_0$.
iii) If $\nu \in \mathcal{M}_G(\mathbb{R})$ and $\delta > 0$, then $\|T_{2\delta}|\nu|\|_0 \leq \|T_\delta|\nu|\|_0$, where $|\nu|$ is the total variation measure of ν.

Proof. i) Since $\int_{\mathbb{R}} p_\delta(x-y)dx = 1$, it follows from Hölder's inequality that

$$\begin{aligned}\|T_\delta\phi\|_0^2 &= \int_{\mathbb{R}}\left|\int_{\mathbb{R}} p_\delta(x-y)\phi(y)dy\right|^2 dx\\ &\leq \int_{\mathbb{R}}\int_{\mathbb{R}} p_\delta(x-y)\phi(y)^2 dydx\\ &= \int_{\mathbb{R}} \phi(y)^2 dy = \|\phi\|_0^2.\end{aligned}$$

ii) Similar to i), we have

$$\begin{aligned}\int_{\mathbb{R}} |T_\delta \nu(x)|^2 dx &= \int_{\mathbb{R}} \left| \int_{\mathbb{R}} p_\delta(x-y)\nu(dy) \right|^2 dx \\ &\le \int_{\mathbb{R}} \int_{\mathbb{R}} p_\delta(x-y)^2 |\nu|(dy)|\nu|(\mathbb{R})dx \\ &= (2\pi\delta)^{-1/2} \left(|\nu|(\mathbb{R})\right)^2 < \infty.\end{aligned}$$

Thus, $T_\delta \nu \in H_0$.

iii) For the simplicity of notation, we assume that $\nu \in \mathcal{M}_F(\mathbb{R})$. Note that

$$\begin{aligned}\|T_{2\delta}\nu\|_0^2 &= \int_{\mathbb{R}} \left| \int_{\mathbb{R}} p_{2\delta}(x-y)\nu(dy) \right|^2 dx \\ &= \int_{\mathbb{R}} \left| \int_{\mathbb{R}} \int_{\mathbb{R}} p_\delta(x-z)p_\delta(y-z)dz\nu(dy) \right|^2 dx \\ &= \int_{\mathbb{R}} \left| \int_{\mathbb{R}} p_\delta(x-z)T_\delta\nu(z)dz \right|^2 dx \\ &\le \int_{\mathbb{R}} \int_{\mathbb{R}} p_\delta(x-z)(T_\delta\nu(z))^2 dzdx = \|T_\delta\nu\|_0^2.\end{aligned}$$

The conclusion follows by taking square roots on both sides. □

Let $Z_s^\delta = T_\delta U_s$ where U_s is an $\mathcal{M}_G(\mathbb{R})$-valued solution to (3.1.1). To obtain an estimate for the H_0-norm of the process Z_t^δ, we adapt Kotelenez's arguments to the present setup (cf. Kotelenez (1992)). Replacing ϕ by $T_\delta\phi$ in (3.1.1) and using the fact that differentiation and T_δ commute, we have

$$\begin{aligned}&\left\langle Z_t^\delta, \phi \right\rangle_0 = \langle U_t, T_\delta\phi\rangle \\ &= \langle U_0, T_\delta\phi\rangle + \int_0^t \langle d_s U_s, T_\delta\phi\rangle\, ds \\ &\quad + \frac{1}{2}\int_0^t \langle U_s, a_s \Delta T_\delta\phi\rangle\, ds + \int_0^t \langle U_s, b_s \nabla T_\delta\phi\rangle\, ds \\ &\quad + \int_0^t \int_U \langle U_s, \alpha_s(\cdot,y)\nabla T_\delta\phi + \beta_s(\cdot,y)T_\delta\phi\rangle\, W(dyds) \\ &= \left\langle Z_0^\delta, \phi\right\rangle_0 + \int_0^t \langle T_\delta(d_s U_s), \phi\rangle_0\, ds \\ &\quad + \frac{1}{2}\int_0^t \langle \Delta T_\delta(a_s U_s), \phi\rangle_0\, ds - \int_0^t \langle \nabla T_\delta(b_s U_s), \phi\rangle_0\, ds \\ &\quad + \int_0^t \int_U \langle -\nabla T_\delta(\alpha_s(\cdot,y)U_s) + T_\delta(\beta_s(\cdot,y)U_s), \phi\rangle_0\, W(dyds).\end{aligned}$$

By Itô's formula, we obtain

$$\begin{aligned}
&\left\langle Z_t^\delta, \phi\right\rangle_0^2 \\
&= \left\langle Z_0^\delta, \phi\right\rangle_0^2 + \int_0^t 2\left\langle Z_s^\delta, \phi\right\rangle_0 \langle T_\delta(d_s U_s), \phi\rangle_0 ds \\
&\quad + \int_0^t \left\langle Z_s^\delta, \phi\right\rangle_0 \langle \Delta T_\delta(a_s U_s), \phi\rangle_0 ds - \int_0^t 2\left\langle Z_s^\delta, \phi\right\rangle_0 \langle \nabla T_\delta(b_s U_s), \phi\rangle_0 ds \\
&\quad + \int_0^t \int_U 2\left\langle Z_s^\delta, \phi\right\rangle_0 \langle -\nabla T_\delta(\alpha_s(\cdot, y)U_s) + T_\delta(\beta_s(\cdot, y)U_s), \phi\rangle_0 W(dyds) \\
&\quad + \int_0^t \int_U |\langle -\nabla T_\delta(\alpha_s(\cdot, y)U_s) + T_\delta(\beta_s(\cdot, y)U_s), \phi\rangle_0|^2 \mu(dy)ds.
\end{aligned}$$

Summing over ϕ in a CONS of H_0 and taking expectations, we have

$$\begin{aligned}
\mathbb{E}\|Z_t^\delta\|_0^2 = \|Z_0^\delta\|_0^2 &+ \mathbb{E}\int_0^t 2\left\langle Z_s^\delta, T_\delta(d_s U_s)\right\rangle_0 ds \\
&+ \mathbb{E}\int_0^t \left\langle Z_s^\delta, \Delta T_\delta(a_s^{(1)} U_s)\right\rangle_0 ds \\
&+ \mathbb{E}\int_0^t \left\langle Z_s^\delta, \Delta T_\delta(a_s^{(2)} U_s)\right\rangle_0 ds \\
&- \mathbb{E}\int_0^t 2\left\langle Z_s^\delta, \nabla T_\delta(b_s U_s)\right\rangle_0 ds \\
&+ \mathbb{E}\int_0^t \int_U \|\nabla T_\delta(\alpha_s(\cdot, y)U_s)\|_0^2 \mu(dy)ds \\
&- 2\mathbb{E}\int_0^t \int_U \langle \nabla T_\delta(\alpha_s(\cdot, y)U_s), T_\delta(\beta_s(\cdot, y)U_s)\rangle_0 \mu(dy)ds \\
&+ \mathbb{E}\int_0^t \int_U \|T_\delta(\beta_s(\cdot, y)U_s)\|_0^2 \mu(dy)ds. \qquad (3.1.3)
\end{aligned}$$

We will show that the integral terms on the right of (3.1.3) are bounded by a constant times the integral of $\|T_\delta(|U_s|)\|_0$.

Lemma 3.1.2. *Let $(H, \mathcal{H}, \eta)$ be a measure space and $\mathbb{H} = L^2(H, \eta)$. (We are interested in the cases H being a singleton and $H = U$ with $\eta = \mu$.) Let $f_i : \mathbb{R} \to \mathbb{H}$, $i = 1, 2$ satisfy*

$$\|f_i(x) - f_i(y)\|_{\mathbb{H}} \le K|x - y|, \qquad \forall x, y \in \mathbb{R}^d,$$

$$\|f_i(x)\|_{\mathbb{H}} \le K, \quad \forall x \in \mathbb{R},$$

$g, \nabla g \in H_0$, and let $\zeta \in \mathcal{M}_G(\mathbb{R})$. Then there exist constants K_1 and K_2 depending on f_1 and f_2 but not on ζ and g such that

$$\|\langle g, f_1 \nabla g\rangle_0\|_{\mathbb{H}} \le \frac{1}{2} K\|g\|_0^2, \qquad (3.1.4)$$

$$\left\| \|T_\delta(f_1\zeta)\|_0 \right\|_{\mathbb{H}} \leq \|T_\delta(\|f_1\|_H |\zeta|)\|_0 \leq K \|T_\delta(|\zeta|)\|_0, \tag{3.1.5}$$

$$\left\| \|f_1 \nabla T_\delta(\zeta) - \nabla T_\delta(f_1\zeta)\|_0 \right\|_{\mathbb{H}} \leq K_1 \|T_{2\delta}(|\zeta|)\|_0, \tag{3.1.6}$$

and

$$\left| \langle T_\delta(f_2\zeta), \nabla T_\delta(f_1\zeta) \rangle_{H_0 \otimes \mathbb{H}} \right| \leq K_2 \|T_\delta(|\zeta|)\|_0, \tag{3.1.7}$$

here notation $H_0 \otimes \mathbb{H}$ stands for the tensor product in general, and in the current special case, it is equal to $L^2(\mathbb{R} \times U, dx\eta(du))$; the notation $T_\delta(f_1\zeta)$ is a mapping from $\mathbb{R}$ to $\mathbb{H}$ given by

$$T_\delta(f_1\zeta)(x) = \int_{\mathbb{R}} p_\delta(x-y) f_1(y)\zeta(dy).$$

Proof. To prove (3.1.4), first assume that f_1 and g are continuously differentiable with compact support. Then integrating by parts we have

$$\begin{aligned} \langle g, f_1(\cdot, u)\nabla g\rangle_0 &= \frac{1}{2}\int_{\mathbb{R}} f_1(x,u)\nabla(g^2(x))dx \\ &= -\frac{1}{2}\int_{\mathbb{R}} g^2(x)\nabla f_1(x,u)dx, \end{aligned} \tag{3.1.8}$$

and hence

$$\begin{aligned} &\int_H |\langle g, f_1(\cdot, u)\nabla g\rangle_0|^2 \eta(du) \\ &= \frac{1}{4}\int_H \left| \int_{\mathbb{R}} g^2(x)\nabla f_1(x,u)dx \right|^2 \eta(du) \\ &\leq \frac{1}{4}\int_H \int_{\mathbb{R}} g^2(x)dx \int_{\mathbb{R}} g^2(x)|\nabla f_1(x,u)|^2 dx\eta(du) \\ &= \frac{1}{4}\|g\|_0^2 \int_H \int_{\mathbb{R}} g^2(x) \lim_{\epsilon\to 0} \epsilon^{-2} |f_1(x+\epsilon,u) - f_1(x,u)|^2 dx\eta(du) \\ &\leq \frac{1}{4}\|g\|_0^2 \int_{\mathbb{R}} g^2(x) \liminf_{\epsilon\to 0} \epsilon^{-2} \int_H |f_1(x+\epsilon,u) - f_1(x,u)|^2 \eta(du)dx \\ &\leq \frac{1}{4}\|g\|_0^2 \int_{\mathbb{R}} g^2(x) \liminf_{\epsilon\to 0} \epsilon^{-2}K^2\epsilon^2 dx \\ &= \frac{1}{4}K^2\|g\|_0^4, \end{aligned}$$

where we used Fatou's lemma for the second inequality and Condition (UB) for the third. Inequality (3.1.4) for f_1 and g not being differentiable follows by approximation.

Writing

$$\begin{aligned}
\| \| T_\delta(f_1\zeta)\|_0\|_{\mathbb{H}}^2 &= \int_H \int_{\mathbb{R}} \left| \int_{\mathbb{R}} p_\delta(x-y) f_1(y,u)\zeta(dy)\right|^2 dx\eta(du) \\
&\le \int_{\mathbb{R}}\int_{\mathbb{R}}\int_{\mathbb{R}} p_\delta(x-y)p_\delta(x-z) \\
&\qquad \times \int_H |f_1(y,u)f_1(z,u)|\eta(du)|\zeta|(dy)|\zeta|(dz)dx \\
&\le \int_{\mathbb{R}} \left(\int_{\mathbb{R}} p_\delta(x-y)\|f_1(y)\|_H |\zeta|(dy)\right)^2 dx,
\end{aligned}$$

(3.1.5) follows.

Note that

$$\begin{aligned}
&\|f_1(x,\cdot)\nabla T_\delta\zeta(x) - \nabla T_\delta(f_1\zeta)(x)\|_{\mathbb{H}} \\
&= \left\| \int_{\mathbb{R}} (f_1(x,\cdot) - f_1(y,\cdot))\nabla p_\delta(x-y)\zeta(dy)\right\|_{\mathbb{H}} \\
&\le \int_{\mathbb{R}} K\frac{|x-y|^2}{\delta} p_\delta(x-y)|\zeta|(dy) \\
&\le K\int_{\mathbb{R}} \frac{|x-y|^2}{\delta}\exp\left(-\frac{|x-y|^2}{4\delta}\right) p_{2\delta}(x-y)|\zeta|(dy) \\
&\le K_1 T_{2\delta}(|\zeta|)(x), \qquad (3.1.9)
\end{aligned}$$

where we used the boundedness of $\{xe^{-x},\ x>0\}$ in the last inequality.

As

$$\| \|h\|_{\mathbb{H}}\|_0 = \| \|h\|_0\|_{\mathbb{H}},$$

taking the H_0-norm of both sides of (3.1.9) gives (3.1.6).

By (3.1.4) and (3.1.6),

$$\begin{aligned}
&\left|\langle T_\delta(f_2\zeta), \nabla T_\delta(f_1\zeta)\rangle_{H_0\otimes\mathbb{H}}\right| \\
&\le \left|\langle T_\delta(f_2\zeta), f_1\nabla T_\delta\zeta\rangle_{H_0\otimes\mathbb{H}}\right| + \left|\langle T_\delta(f_2\zeta), \nabla T_\delta(f_1\zeta) - f_1\nabla T_\delta(\zeta)\rangle_{H_0\otimes\mathbb{H}}\right| \\
&\le \left|\langle T_\delta(f_2\zeta), \nabla f_1 T_\delta(\zeta)\rangle_{H_0\otimes\mathbb{H}}\right| + \left|\langle f_2\nabla T_\delta\zeta, f_1 T_\delta(\zeta)\rangle_{H_0\otimes\mathbb{H}}\right| \\
&\quad + \left|\langle(\nabla T_\delta(f_2\zeta) - f_2\nabla T_\delta\zeta), f_1 T_\delta\zeta\rangle_{H_0\otimes\mathbb{H}}\right| \\
&\quad + K_1\|T_\delta(f_2\zeta)\|_{H_0\otimes\mathbb{H}}\|T_{2\delta}(|\zeta|)\|_0 \\
&\le K\| \|T_\delta(f_2\zeta)\|_0\|_{\mathbb{H}}\|T_\delta\zeta\|_0 + K^2\|T_\delta(|\zeta|)\|_0^2 \\
&\quad + KK_1\|T_\delta(|\zeta|)\|_0^2 + K_1K\|T_\delta(|\zeta|)\|_0^2 \\
&\le K_3\|T_\delta(|\zeta|)\|_0^2,
\end{aligned}$$

where the next to the last inequality follows by Lemma 3.1.1 iii) and the previous inequalities. □

Lemma 3.1.3. *Let $\alpha : \mathbb{R} \to \mathbb{H} = L^2(H, \eta)$ satisfy*

$$\|\alpha(x)\|_{\mathbb{H}} \le K \text{ and } \|\alpha(x) - \alpha(y)\|_{\mathbb{H}} \le K|x-y|, \qquad \forall\, x, y \in \mathbb{R},$$

where K is a constant. Define

$$a(x) = \int_H \alpha(x, y)\alpha(x, y)\eta(dy), \qquad \forall\, x \in \mathbb{R}.$$

Then, there exists a constant K_1 such that for $\zeta \in \mathcal{M}_G(\mathbb{R})$,

$$\langle T_\delta \zeta, \Delta T_\delta(a\zeta)\rangle_0 + \int_H \|\nabla T_\delta(\alpha(\cdot, u)\zeta)\|_0^2\, \eta(du) \le K_1 \, \|T_\delta|\zeta|\|_0^2 . \tag{3.1.10}$$

Proof. Notice that

$$\begin{aligned}
&\langle T_\delta \zeta, \Delta T_\delta(a\zeta)\rangle_0 \\
&= \int_{\mathbb{R}} dx \int_{\mathbb{R}} p_\delta(x-y)\zeta(dy) \int_{\mathbb{R}} \Delta p_\delta(x-z)a(z)\zeta(dz) \\
&= \int_{\mathbb{R}} \zeta(dy) \int_{\mathbb{R}} a(z)\zeta(dz)\Delta_z \int_{\mathbb{R}} p_\delta(x-y)p_\delta(x-z)dx \\
&= \int_{\mathbb{R}} \zeta(dy) \int_{\mathbb{R}} a(z)\zeta(dz)\Delta_z p_{2\delta}(y-z) \\
&= \int_{\mathbb{R}} \zeta(dy) \int_{\mathbb{R}} \zeta(dz) \left(\frac{|z-y|^2}{4\delta^2} - \frac{1}{2\delta}\right) p_{2\delta}(z-y)a(z) \\
&= \int_{\mathbb{R}} \zeta(dy) \int_{\mathbb{R}} \zeta(dz) \left(\frac{|z-y|^2}{4\delta^2} - \frac{1}{2\delta}\right) p_{2\delta}(z-y)\frac{1}{2}(a(z) + a(y))
\end{aligned}$$

where the last equality follows from the symmetry in y, z. Similarly,

$$\begin{aligned}
&\int_H \|\nabla T_\delta(\alpha(\cdot, u)\zeta)\|_0^2\, \eta(du) \\
&= -\int_H \eta(du)\, \langle T_\delta(\alpha(\cdot, u)\zeta), \Delta T_\delta(\alpha(\cdot, u)\zeta)\rangle_0 \\
&= -\int_H \eta(du) \int_{\mathbb{R}} \zeta(dy) \int_{\mathbb{R}} \zeta(dz) \left(\frac{|z-y|^2}{4\delta^2} - \frac{1}{2\delta}\right) \\
&\qquad \times p_{2\delta}(z-y)\alpha(y, u)\alpha(z, u).
\end{aligned}$$

Hence, we can estimate the left hand side (LHS) of (3.1.10) as

$$\begin{aligned}
&\text{LHS of (3.1.10)}\\
&= \int_{\mathbb{R}} \zeta(dy) \int_{\mathbb{R}} \zeta(dz) \left(\frac{|y-z|^2}{4\delta^2} - \frac{1}{2\delta} \right) p_{2\delta}(z-y)\\
&\qquad \times \frac{1}{2} \int_H (\alpha(y,u) - \alpha(z,u))^2 \eta(du)\\
&\le \frac{1}{2} \int_{\mathbb{R}} |\zeta|(dy) \int_{\mathbb{R}} |\zeta|(dz) \left(\frac{|z-y|^2}{4\delta^2} + \frac{1}{2\delta} \right) \exp\left(-\frac{|z-y|^2}{4\delta} \right)\\
&\qquad \times \sqrt{2} p_{4\delta}(z-y) K^2 |z-y|^2\\
&\le 16K^2 \int_{\mathbb{R}} |\zeta|(dy) \int_{\mathbb{R}} |\zeta|(dz) \sqrt{2} p_{4\delta}(z-y)\\
&= 16\sqrt{2}K^2 \left\| T_{2\delta} |\zeta| \right\|_0^2\\
&\le 16\sqrt{2}K^2 \left\| T_{\delta} |\zeta| \right\|_0^2,
\end{aligned}$$

where the second inequality follows by bounding $(v^2+v)e^{-v/4}$. The lemma follows with $K_1 = 16\sqrt{2}K^2$. □

The estimates in Lemmas 3.1.2 and 3.1.3 give the following.

Theorem 3.1.4. *If U is an $\mathcal{M}_G(\mathbb{R})$-valued solution of (3.1.1) and $Z^\delta = T_\delta U$, then*

$$\mathbb{E}\|Z_t^\delta\|_0^2 \le \|Z_0^\delta\|_0^2 + K_1 \int_0^t \mathbb{E}\left\| T_\delta |U_s| \right\|_0^2 ds \tag{3.1.11}$$

where K_1 is a constant.

Proof. Using (3.1.5), the second and last terms on the right hand side (RHS) of (3.1.3) are bounded by a constant times $\|T_\delta |U_s|\|_0^2$. The bound for the third term follows from Lemma 3.1.3 (take H to be a singleton). The bound for the sum of the fourth and sixth terms also follows from Lemma 3.1.3 (take $\mathbb{H} = L^2(U,\mu)$). Furthermore, the bound for the fifth and seventh terms follows by (3.1.7). □

Corollary 3.1.5. *If U is an $\mathcal{M}_F(\mathbb{R})$-valued solution of (3.1.1) and $U_0 \in H_0$, then $U_t \in H_0$ a.s. and $\mathbb{E}\|U_t\|_0^2 < \infty$, $\forall t \ge 0$.*

Proof. It follows from (3.1.11) that

$$\mathbb{E}\|Z_t^\delta\|_0^2 \le \|Z_0^\delta\|_0^2 + K_1 \int_0^t \mathbb{E}\|Z_s^\delta\|_0^2 ds.$$

By Gronwall's inequality, we have

$$\mathbb{E}\|Z_t^\delta\|_0^2 \le \|Z_0^\delta\|_0^2 e^{K_1 t}.$$

Let $\{\phi_j\}$ be a CONS of H_0 such that $\phi_j \in C_b(\mathbb{R})$. Then, by Fatou's lemma,

$$\begin{aligned}\mathbb{E}\left[\sum_j \langle U_t, \phi_j\rangle^2\right] &= \mathbb{E}\left[\sum_j \lim_{\delta\to 0} \langle U_t, \phi_j\rangle^2\right] \\ &\le \liminf_{\delta\to 0} \mathbb{E}\|Z_t^\delta\|_0^2 \le \|U_0\|_0^2 e^{K_6 t}.\end{aligned}$$

Hence $U_t \in H_0$ and $\mathbb{E}\|U_t\|_0^2 < \infty$. □

These estimates give uniqueness of $\mathcal{M}_F(\mathbb{R})$-valued solutions with $U_0 \in H_0$.

Theorem 3.1.6. *Suppose that $U_0 \in H_0$ such that $U_0 \ge 0$. Then (3.1.1) has at most one $\mathcal{M}_F(\mathbb{R})$-valued solution.*

Proof. Let U_t^1 and U_t^2 be two $\mathcal{M}_F(\mathbb{R})$-valued solutions with the same initial value U_0. By Corollary 3.1.5, U_t^1, $U_t^2 \in H_0$ a.s. Let $U_t = U_t^1 - U_t^2$. Then $U_t \in H_0$ and

$$\mathbb{E}\|T_\delta U_t\|_0^2 \le K_1 \int_0^t \mathbb{E}\,\|T_\delta|U_s|\|_0^2\, ds.$$

As before, taking $\delta \to 0$, we have

$$\mathbb{E}\|U_t\|_0^2 \le K_1 \int_0^t \mathbb{E}\,\||U_s|\|_0^2\, ds = K_1 \int_0^t \mathbb{E}\,\|U_s\|_0^2\, ds, \tag{3.1.12}$$

and by Gronwall's inequality, we obtain $U_t \equiv 0$. □

By exactly the same argument we have the following theorem.

Theorem 3.1.7. *Suppose that $U_0 \in H_0$. Then (3.1.1) has at most one H_0-valued solution.*

Proof. Let U_t^1 and U_t^2 be two H_0-valued solutions with the same initial value U_0. Let $U_t = U_t^1 - U_t^2$. Then $U_t \in H_0$ and

$$\mathbb{E}\|T_\delta U_t\|_0^2 \le K_1 \int_0^t \mathbb{E}\,\|T_\delta|U_s|\|_0^2\, ds.$$

Taking $\delta \to 0$, we have

$$\mathbb{E}\|U_t\|_0^2 \le K_1 \int_0^t \mathbb{E}\,\||U_s|\|_0^2\, ds = K_1 \int_0^t \mathbb{E}\,\|U_s\|_0^2\, ds.$$

By Gronwall's inequality, we have $U_t \equiv 0$. □

3.2 A duality representation

In this section, we provide a duality representation of the process X_t determined by SPDE (3.0.1). This representation will be useful in calculating the conditional Laplace transform of the superprocess in a random environment in Chapter 5, as well as in the proof of the Hölder continuity of the density random field for the SPRE in Chapter 6.

To aid the understanding of this method, we recall the duality used in the proof of uniqueness for the solution of a linear partial differential equation (PDE). Let u be a solution to the following PDE:

$$\frac{\partial u}{\partial s} = L^* u \tag{3.2.1}$$

with initial condition u_0, where L is a second order differential operator and L^* is the adjoint operator of L. To prove the uniqueness for the solution of (3.2.1), we consider the following backward PDE for $s \in [0,t]$ with t being fixed:

$$\begin{cases} \frac{\partial v}{\partial s} = -Lv, \\ v_t = g. \end{cases} \tag{3.2.2}$$

Then $\frac{d}{ds}\langle u_s, v_s\rangle_0 = 0$ and, hence,

$$\langle u_t, g\rangle_0 = \langle u_0, v_0\rangle_0 ,$$

where notation $\langle\cdot,\cdot\rangle_0$ is the inner product in $L^2(\mathbb{R})$. This implies the uniqueness for the solution to (3.2.1).

Now we imitate (3.2.2) and consider the backward SPDE:

$$u_{r,t}(x) = f(x) + \int_r^t Lu_{s,t}(x)ds + \int_r^t \int_U \nabla u_{s,t}(x)h(y,x)W(\hat{d}sdy), \tag{3.2.3}$$

where $\hat{d}s$ stands for the backward Itô integral. Namely, $u_{s,t}(x)$ is measurable with respect to the σ-field

$$\mathcal{F}^W_{s,t} = \sigma\left(W([r,t)\times A) : r \in [s,t],\ A \in \mathcal{U}\right),$$

and the stochastic integral is approximated as

$$\int_r^t \int_U \nabla u_{s,t}(x)W(\hat{d}sdy) = \lim_{n\to\infty}\sum_{i=1}^n \int_U \nabla u_{s_i,t}(x)h(y,x)W([s_{i-1},s_i)\times dy),$$

where $r = s_0 < s_1 < \cdots < s_n = t$ is a partition with

$$\max_{1\le i\le n}(s_i - s_{i-1}) \to 0.$$

We also denote $u_{r,t}(x)$ by $u^f_{r,t}(x)$ to indicate its dependence on f.

We now establish the duality relationship between (X_t) and $(u^f_{0,t})$ given by the SPDEs (3.0.1) and (3.2.3), respectively. To begin with, we introduce some notations needed. Recall that $H_0 = L^2(\mathbb{R})$ is the set of all square integrable functions on $\mathbb{R}$, and H_0^+ consists of all the nonnegative functions in H_0. Denote

$$H_m = \{\phi \in H_0 : \phi^{(j)} \in H_0, \;\; j \leq n\},$$

where $\phi^{(j)}$ stands for the jth derivative of ϕ in the sense of generalized functions. We refer the reader to Section 2.1 of Chapter 1 in the book of Gel'fand and Shilov (1964) for a precise definition of such derivatives. In this section, we will use H_n for integer n only. Therefore, we will use the classical definition of the Sobolev norms which is equivalent to those defined in the next section. Namely, we define Sobolev norm on H_m by

$$\|\phi\|_m^2 = \sum_{j=0}^m \int_{\mathbb{R}} |\phi^{(j)}(x)|^2 dx.$$

To prove the differentiability for the solution, we now convert the backward SPDE (3.2.3) to an ordinary SPDE by reversing the time parameter. Fix $t > 0$. For $0 < s < t$, we define

$$\mathbb{W}([0,s] \times A) = W((t-s,t] \times A) \quad \text{and} \quad v_s = u_{t-s,t}.$$

Then v_s satisfies the following forward SPDE

$$v_s(x) = f(x) + \int_0^s Lv_r(x)ds + \int_0^s \int_U \nabla v_r(x)h(y,x)\mathbb{W}(drdy). \tag{3.2.4}$$

The uniqueness for the solution to (3.2.4) with $f \in H_0$ follows from Theorem 3.1.7. However we need here the solution of (3.2.3) to be a process with values in $C_b^2(\mathbb{R})$. To achieve this, we show that $v_s \in H_k$, where k is chosen so that $2(k-2) > 1$, and then, using a standard Sobolev imbedding argument, which we state below (without giving the proof) for the convenience of the reader. We refer the reader to the book of Adams (1975) for the proof of a more general version of the theorem.

Theorem 3.2.1 (Sobolev). *If $2(k-j) > 1$, then H_k can be embedded into $C_b^j(\mathbb{R})$, i.e., there is a constant K_1 and a linear mapping from $f \in H_k$ to $\bar{f} \in C_b^j(\mathbb{R})$ such that $f(x) = \bar{f}(x)$ for almost every x and*

$$\|\bar{f}\|_{j,\infty} \equiv \sup_{x\in\mathbb{R}} \sum_{k=0}^j |f^{(i)}(x)| \leq K_1 \|f\|_k.$$

Now we are ready to prove the existence of a smooth solution to SPDE (3.2.4). We shall need the following:

Assumption (BD): The mappings $h \in C_b^3(\mathbb{R}, L^2(U,\mu))$ and $a, f \in C_b^3(\mathbb{R})$, where the notation $C_b^3(\mathbb{R}, \mathcal{X})$ denotes all bounded continuous mappings from $\mathbb{R}$ to $\mathcal{X}$ with bounded continuous derivatives up to order 3 ($\mathcal{X}$ is omitted if it equals $\mathbb{R}$). Also, we assume $f \in H_3$.

Lemma 3.2.2. *Suppose that Assumption (BD) holds. Then there exists a constant K_1 independent of f and $s \in [0,t]$ such that*

$$\mathbb{E}[\|v_s\|_3^2] \le K_1 \|f\|_3^2. \tag{3.2.5}$$

As a consequence $v_s \in C_b^2(\mathbb{R})$, a.s. and there exists a constant K_2 independent of f and $s \in [0,t]$ such that

$$\mathbb{E}[\|v_s\|_{2,\infty}^2] \le K_2 \|f\|_3^2. \tag{3.2.6}$$

Proof. It follows from the same arguments as those leading to (3.1.12) that there exists a constant K_3 such that

$$\mathbb{E}\|v_s\|_0^2 \le K_3 \|f\|_0^2. \tag{3.2.7}$$

Next, we take derivative (smooth out by the Brownian semigroup T_δ as we did in Section 3.1 if necessary) on both sides of (3.2.4). Then, $v_s^1 \equiv \nabla v_s$ satisfies the following SPDE

$$v_s^1(x) = f'(x) + \int_0^s L_2 v_r^1(x) ds + \int_0^s \int_U L_{1,y} v_r^1(x) \mathbb{W}(drdy),$$

where

$$L_{1,y} f(x) = \nabla h(y,x) f(x) + h(y,x) f'(x)$$

and

$$L_2 f(x) = \frac{1}{2} a(x) f''(x) + \frac{1}{2} a'(x) f'(x).$$

Similar to (3.2.7), we can prove that

$$\mathbb{E}\|v_s^1\|_0^2 \le K_4 \|f'\|_0^2.$$

Same arguments apply to 2nd and 3rd derivatives we finish the proof of (3.2.5). The estimate (3.2.6) then follows from Sobolev's imbedding theorem.

□

For $\phi \in L^2([0,t] \times U, ds\mu(du))$, we define

$$\theta_\phi^W(r) = \exp\left(\sqrt{-1}\int_0^r\int_U \phi_s(u)W(dsdu) + \frac{1}{2}\int_0^r\int_U |\phi_s(u)|^2\mu(du)ds\right). \tag{3.2.8}$$

We say that ϕ is bounded in $L^2(U,\mu)$ if

$$\sup_{s\le t}\|\phi_s\|_{L^2(U,\mu)} < \infty.$$

We will need the following lemma which implies that the family of random variables

$$\{\theta_\phi^W(t) : \ \phi \text{ is bounded in } L^2(U,\mu)\}$$

is dense in $L^2(\Omega, \mathcal{F}_t^W, P)$.

Lemma 3.2.3. *If $\xi \in L^2(\Omega, \mathcal{F}_t^W, P)$ satisfies*

$$\mathbb{E}\left(\xi\theta_\phi^W(t)\right) = 0,$$

for all $L^2(U,\mu)$-bounded functions ϕ on $[0,t]$, then $\xi = 0$ a.s.

Proof. Let $\{A_i : \ i = 1, 2, \cdots\}$ be a countable family of subsets of U generating $\mathcal{U}$. Define

$$\mathcal{H}_n = \sigma\left(W((t_{i-1}^n, t_i^n] \times A_j) : \ 1 \le i \le 2^n, 1 \le j \le n\right),$$

where $t_i^n = it2^{-n}, \ \ i = 0, 1, 2, \cdots, 2^n$. Then $\{\mathcal{H}_n\}$ is a sequence of σ-fields increasing to $\mathcal{F}_t^W$. By martingale convergence theorem, we have

$$\xi_n \equiv \mathbb{E}(\xi|\mathcal{H}_n) \to \xi, \qquad \text{a.s.}$$

Note that ξ_n is a function of random variables

$$\left(W((t_{i-1}^n, t_i^n] \times A_j) : \ 1 \le i \le 2^n, 1 \le j \le n\right).$$

Let

$$\phi_s^n(u) = \sum_{i=1}^{2^n}\sum_{j=1}^{n} \lambda_{ij} 1_{[t_{i-1}^n, t_i^n)}(s) 1_{A_j}(u),$$

where λ_{ij}'s are constants. Then

$$0 = \mathbb{E}\left(\xi\theta_{\phi^n}^W(t)\right) = \mathbb{E}\left(\mathbb{E}\left(\xi\theta_{\phi^n}^W(t)|\mathcal{H}_n\right)\right) = \mathbb{E}\left(\xi_n\theta_{\phi^n}^W(t)\right).$$

Note that $\int_0^t\int_U |\phi_s^n(u)|^2\mu(du)ds$ is non-random. This implies that the Fourier transformation of ξ_n is

$$\mathbb{E}\left(\xi_n \exp\left(\sqrt{-1}\sum_{i=1}^{2^n}\sum_{j=1}^{n}\lambda_{ij}W((t_{i-1}^n, t_i^n] \times A_j)\right)\right) = 0.$$

Therefore $\xi_n = 0$ a.s. and hence, $\xi = 0$ a.s. □

The following lemma will play a key role in the proof of the duality representation of the process X_t. Let ξ_t be the process describing the movement of one particle in the system. Namely, ξ_t is governed by the following SDE:

$$d\xi_t = dB_t + \int_U h(y, \xi_t) W(dtdy), \tag{3.2.9}$$

where (B_t) is a Brownian motion independent of W.

Denote $u_{s,t}$ by u_s for the simplicity of notation. Note that u_s is $\mathcal{F}_t^W$-measurable which is independent of $\mathcal{F}_r^B$. The stochastic integral $\int_0^r \nabla u_s(\xi_s) dB_s$ is well-defined on the stochastic basis $(\Omega, \mathcal{F}, P, \tilde{\mathcal{F}}_s)$, where $\tilde{\mathcal{F}}_s = \mathcal{F}_s \vee \mathcal{F}_t^W$, $0 \le s \le t$.

Lemma 3.2.4. *Suppose that Condition (BD) holds. Then, for every $t \ge 0$, we have*

$$u_t(\xi_t) - u_0(\xi_0) = \int_0^t \nabla u_s(\xi_s) dB_s, \qquad a.s.. \tag{3.2.10}$$

Proof. Let ϕ and ψ be two bounded smooth functions on $[0, t]$ taking values in $L^2(U, \mu)$ and $\mathbb{R}$, respectively. Let $\theta_\phi^W(r)$ be defined as in (3.2.8), and let $\theta_\psi^B(r)$ be defined in a similar fashion. Note that both sides of (3.2.10) are $\mathcal{F}_t^W \vee \mathcal{F}_t^B$-measurable. It follows from the previous lemma (with W replaced by (B, W)) that in order to prove (3.2.10) it is sufficient to show that for all bounded functions ϕ and ψ, we have

$$\begin{aligned} &\mathbb{E}\left((u_t(\xi_t) - u_0(\xi_0)) \, \theta_\phi^W(t) \theta_\psi^B(t) \right) \\ &= \mathbb{E}\left(\int_0^t \nabla u_s(\xi_s) dB_s \theta_\phi^W(t) \theta_\psi^B(t) \right). \end{aligned} \tag{3.2.11}$$

Let

$$\Theta_r(x) = \mathbb{E}\left(u_r(x) \tilde{\theta}_\phi^W(r) | \mathcal{F}_r \right), \qquad \forall x \in \mathbb{R}^d$$

where

$$\begin{aligned} &\tilde{\theta}_\phi^W(r) = \theta_\phi^W(t)/\theta_\phi^W(r) \\ &= \exp\left(\sqrt{-1} \int_r^t \int_U \phi_s(y) W(dsdy) + \frac{1}{2} \int_r^t \int_U |\phi_s(y)|^2 \mu(dy) ds \right). \end{aligned}$$

Let $\tilde{\theta}_\psi^B(r)$ be defined similarly.

Since u_r and $\tilde{\theta}_\phi^W(r)$ are measurable with respect to the σ-field

$$\mathcal{F}_{r,t}^W = \sigma\left(W([s,t] \times A) : \; s \le t, \; A \in \mathcal{U} \right),$$

which is independent of $\mathcal{F}_r$, we get that

$$\Theta_r(x) = \mathbb{E}\left(u_r(x)\tilde{\theta}^W_\phi(r)\right).$$

As $\tilde{\theta}^B_\psi(r)$ is independent of $\mathcal{F}_r \vee \mathcal{F}^W_{r,t}$ and $\theta^B_\psi(r)$ is a martingale, we have

$$\begin{aligned}
\mathbb{E}\left(u_r(x)\tilde{\theta}^W_\phi(r)\tilde{\theta}^B_\psi(r)|\mathcal{F}_r\right) &= \mathbb{E}\left(\mathbb{E}\left(u_r(x)\tilde{\theta}^W_\phi(r)\tilde{\theta}^B_\psi(r)\Big|\mathcal{F}_r \vee \mathcal{F}^W_{r,t}\right)\Big|\mathcal{F}_r\right)\\
&= \mathbb{E}\left(u_r(x)\tilde{\theta}^W_\phi(r)\mathbb{E}\left(\tilde{\theta}^B_\psi(r)\Big|\mathcal{F}_r\right)\Big|\mathcal{F}_r\right)\\
&= \mathbb{E}\left(u_r(x)\tilde{\theta}^W_\phi(r)\Big|\mathcal{F}_r\right)\\
&= \Theta_r(x).
\end{aligned}$$

Hence, for $r \in [0,t]$, we have

$$\begin{aligned}
\mathbb{E}\left(u_r(\xi_r)\theta^W_\phi(t)\theta^B_\psi(t)|\mathcal{F}_r\right) &= \theta^W_\phi(r)\theta^B_\psi(r)\mathbb{E}\left(u_r(\xi_r)\tilde{\theta}^W_\phi(r)\tilde{\theta}^B_\psi(r)|\mathcal{F}_r\right)\\
&= \Theta_r(\xi_r)\theta^W_\phi(r)\theta^B_\psi(r). \qquad (3.2.12)
\end{aligned}$$

Note that $\int_r^t \int_U \phi_s(y)W(\hat{d}sdy)$ coincides with $\int_r^t \int_U \phi_s(y)W(dsdy)$ since $\phi_s(y)$ is deterministic. Thus, by the backward Itô's formula, we have

$$\hat{d}\tilde{\theta}^W_\phi(r) = -\sqrt{-1}\tilde{\theta}^W_\phi(r)\int_U \phi_r(y)W(\hat{d}rdy). \qquad (3.2.13)$$

Applying the backward Itô's formula to (3.2.3) and (3.2.13), we get

$$\begin{aligned}
\hat{d}(u_r\tilde{\theta}^W_\phi(r)) =& -Lu_r\tilde{\theta}^W_\phi(r)dr - \int_U \nabla u_r h(y,\cdot)\tilde{\theta}^W_\phi(r)W(\hat{d}rdy)\\
&- \sqrt{-1}\int_U u_r\tilde{\theta}^W_\phi(r)\phi_r(y)W(\hat{d}rdy)\\
&+ \sqrt{-1}\int_U \nabla u_r h(y,\cdot)\phi_r(y)\tilde{\theta}^W_\phi(r)\mu(dy)dr\\
=& \left(-Lu_r + \sqrt{-1}\int_U \nabla u_r h(y,\cdot)\phi_r(y)\mu(dy)\right)\tilde{\theta}^W_\phi(r)dr\\
&- \int_U \left(\nabla u_r h(y,\cdot) - \sqrt{-1}u_r\phi(r,y)\right)\tilde{\theta}^W_\phi(r)W(\hat{d}rdy).
\end{aligned}$$

Writing into integral form, we get

$$\begin{aligned}
u_s\tilde{\theta}^W_\phi(s) - f =& \int_s^t \left(Lu_r - \sqrt{-1}\int_U \nabla u_r h(y,\cdot)\phi_r(y)\mu(dy)\right)\tilde{\theta}^W_\phi(r)dr\\
&+ \int_U \left(\nabla u_r h(y,\cdot) - \sqrt{-1}u_r\phi(r,y)\right)\tilde{\theta}^W_\phi(r)W(\hat{d}rdy).
\end{aligned}$$

Taking expectation on both sides, we see that

$$\Theta_s - f = \int_s^t \left(L\Theta_r - \sqrt{-1} \int_U \nabla\Theta_r h(y,\cdot)\phi_r(y)\mu(dy) \right) dr,$$

and hence, Θ_r is the solution to the following PDE:

$$\frac{d}{dr}\Theta_r = -L\Theta_r + \sqrt{-1} \int_U \nabla\Theta_r h(y,\cdot)\phi_r(y)\mu(dy).$$

As a consequence, Θ is differentiable in r and has continuous first and second order partial derivatives in x.

By Itô's formula, we have

$$\begin{aligned} d\Theta_r(\xi_r) &= \left(-L\Theta_r(\xi_r) + \sqrt{-1} \int_U \nabla\Theta_r(\xi_r) h(y,\xi_r)\phi_r(y)\mu(dy) \right) dr \\ &\quad + L\Theta_r(\xi_r)dr + \nabla\Theta_r(\xi_r) \left(dB_r + \int_U h(y,\xi_r) W(drdy) \right) \\ &= \sqrt{-1} \int_U \nabla\Theta_r(\xi_r) h(y,\xi_r)\phi_r(y)\mu(dy)dr \\ &\quad + \nabla\Theta_r(\xi_r) \left(dB_r + \int_U h(y,\xi_r) W(drdy) \right). \end{aligned} \tag{3.2.14}$$

Note that

$$d\theta_\phi^W(r) = \sqrt{-1}\theta_\phi^W(r) \int_U \phi_s(y) W(drdy)$$

and

$$d\theta_\psi^B(r) = \sqrt{-1}\theta_\psi^B(r)\psi_r dB_r.$$

Applying Itô's formula to the three equations above, we get

$$\begin{aligned} &d(\Theta_r(\xi_r)\theta_\phi^W(r)\theta_\psi^B(r)) \\ &= \sqrt{-1}\theta_\phi^W(r)\theta_\psi^B(r)\nabla\Theta_r(\xi_r)\psi_r dr + d(mart.). \end{aligned} \tag{3.2.15}$$

Making use of (3.2.12) with $r = t$ and $r = 0$, respectively, we get

$$\begin{aligned} &\mathbb{E}\left((u_t(\xi_t) - u_0(\xi_0))\, \theta_\phi^W(t)\theta_\psi^B(t) \right) \\ &= \mathbb{E}\left(\mathbb{E}\left(u_t(\xi_t)\theta_\phi^W(t)\theta_\psi^B(t) \middle| \mathcal{F}_t \right) \right) - \mathbb{E}\left(u_0(\xi_0)\theta_\phi^W(t)\theta_\psi^B(t) \right) \\ &= \mathbb{E}\left(\Theta_t(\xi_t)\theta_\phi^W(t)\theta_\psi^B(t) \right) - \Theta_0(\xi_0) \\ &= \sqrt{-1} \int_0^t \mathbb{E}\left(\theta_\phi^W(r)\theta_\psi^B(r)\nabla\Theta_r(\xi_r)\psi_r \right) dr, \end{aligned} \tag{3.2.16}$$

where the last equality follows from (3.2.15).

Applying Itô's formula on the stochastic basis $(\Omega, \mathcal{F}, P, \tilde{\mathcal{F}}_s)$, we get

$$\begin{aligned}&\int_0^r \nabla u_s(\xi_s)dB_s\theta_\psi^B(r)\\&= \int_0^r \cdots dB_s + \sqrt{-1}\int_0^r \nabla u_s(\xi_s)\psi_s(\xi_s)\theta_\psi^B(s)ds.\end{aligned}$$

This implies that

$$\begin{aligned}&\mathbb{E}\left(\int_0^t \nabla u_s(\xi_s)dB_s\theta_\phi^W(t)\theta_\psi^B(t)\right)\\&= \mathbb{E}\left(\mathbb{E}\left(\int_0^t \nabla u_s(\xi_s)dB_s\theta_\phi^W(t)\theta_\psi^B(t)\Big|\mathcal{F}_t^W\right)\right)\\&= \mathbb{E}\left(\sqrt{-1}\int_0^t \nabla u_s(\xi_s)\psi_s(\xi_s)\theta_\psi^B(s)ds\theta_\phi^W(t)\right)\\&= \sqrt{-1}\int_0^t \mathbb{E}\left(\mathbb{E}\left(\nabla u_s(\xi_s)\psi_s(\xi_s)\theta_\psi^B(s)\theta_\phi^W(t)|\mathcal{F}_s\right)\right)ds\\&= \mathbb{E}\Bigg(\sqrt{-1}\int_0^t \mathbb{E}\left(\nabla u_s(\xi_s)\tilde{\theta}_\phi^W(s)|\mathcal{F}_s\right)\psi_s(\xi_s)\theta_\psi^B(s)\theta_\phi^W(s)ds\Bigg)\\&= \mathbb{E}\left(\sqrt{-1}\int_0^t \nabla\Theta_s(\xi_s)\psi_s(\xi_s)\theta_\psi^B(s)\theta_\phi^W(s)ds\right).\end{aligned} \tag{3.2.17}$$

From (3.2.16) and (3.2.17), we see that (3.2.11) holds, which then implies (3.2.10). □

Theorem 3.2.5. *Suppose that $f \in C_b(\mathbb{R})$ and $h \in C_b\left(\mathbb{R}, L^2(U,\mu)\right)$. Then,*

$$\langle X_t, f\rangle = \left\langle \nu, u_{0,t}^f \right\rangle, \qquad a.s. \tag{3.2.18}$$

Proof. Note that (3.0.1) is a special case of the Zakai equation in the nonlinear filtering theory (cf. Xiong (2008)) with state equation (3.2.9) and observation W. As the observation function is 0, Zakai equation is satisfied by the optimal filter π_t given by

$$\pi_t f \equiv \mathbb{E}\left(f(\xi_t)|\mathcal{F}_t^W\right), \qquad \forall\, f \in C_b(\mathbb{R}).$$

Thus, by the uniqueness for the solution to (3.0.1), we see that

$$\langle X_t, f\rangle = \mathbb{E}\left(f(\xi_t)\big|\mathcal{F}_t^W\right).$$

Next, we assume that Condition (BD) holds. By independency of W and B, we see that

$$\mathbb{E}\left(\int_0^t \nabla u_s(\xi_s)dB_s\Big|\mathcal{F}_t^W\right) = 0.$$

By (3.2.10), we then have

$$\langle X_t, f\rangle = \mathbb{E}\left(f(\xi_t)\middle|\mathcal{F}_t^W\right) = \left\langle \nu, u_{0,t}^f\right\rangle.$$

Now we remove Condition (BD). Let $\{(h^n, f^n)\}$ be a sequence of functions such that for each n, Condition (BD) is satisfied, and as $n \to \infty$, it converges to (h, f) in the following sense:

$$\sup_{x\in\mathbb{R}}\left(|f^n(x) - f(x)|^2 + \int_U |h^n(y,x) - h(y,x)|^2\mu(dy)\right) \to 0.$$

By the proof above, we have that

$$\langle X_t^n, f\rangle = \langle \nu, u_0^n\rangle, \tag{3.2.19}$$

where (X_s^n) and (u_s^n) are the solutions of SPDEs (3.0.1) and (3.2.3), respectively, with (f, h) replaced by (f^n, h^n).

Let ξ_t^n be the solution of the equation

$$d\xi_s^n = dB_s + \int_U h^n(y, \xi_s^n)W(dsdy).$$

By Burkholder-Davis-Gundy inequality, we get

$$\begin{aligned}\mathbb{E}\sup_{r\le s}|\xi_r^n - \xi_r|^2 \le 8t\sup_{x\in\mathbb{R}}\int_U |h^n(y,x) - h(y,x)|^2\mu(dy)\\ + 8K\int_r^s \mathbb{E}|\xi_r^n - \xi_r|^2 dr.\end{aligned}$$

Gronwall's inequality then implies

$$\mathbb{E}\sup_{r\le t}|\xi_r^n - \xi_r|^2 \to 0.$$

Thus

$$\begin{aligned}\mathbb{E}\left|\langle X_t^n, f^n\rangle - \langle X_t, f\rangle\right| &= \mathbb{E}\left|\mathbb{E}^W\left(f^n(\xi_t^n) - f(\xi_t)\right)\right|\\ &\le \sup_{x\in\mathbb{R}}|f^n(x) - f(x)| + \mathbb{E}|f(\xi_t^n) - f(\xi_t)|\\ &\to 0.\end{aligned}$$

Similarly, we can prove the convergence of u_0^n to u_0. The conclusion then follows by taking $n \to \infty$ on both sides of (3.2.19).

□

As a consequence of the duality relation (3.2.18), we have the following corollary which will be useful in Chapter 5.

Corollary 3.2.6. *For any $f \in C_b^2(\mathbb{R})$, we have*

$$u_{0,t}^f = f + \int_0^t u_{0,s}^{Lf} ds + \int_0^t\int_U u_{0,s}^{h(y,\cdot)f'} W(dsdy). \tag{3.2.20}$$

Proof. Note that for $f \in C_b^2(\mathbb{R})$ we have

$$\begin{aligned}\left\langle \nu, u_{0,t}^f \right\rangle &= \mathbb{E}^W \langle X_t, f\rangle \\ &= \langle \nu, f\rangle + \int_0^t \mathbb{E}^W \langle X_s, Lf\rangle\, ds - \int_0^t \int_U \mathbb{E}^W \langle X_s, h(y,\cdot)f'\rangle\, W(dsdy) \\ &= \left\langle \nu, f + \int_0^t u_{0,s}^{Lf} ds + \int_0^t \int_U u_{0,s}^{h(y,\cdot)f'} W(dsdy) \right\rangle.\end{aligned}$$

This implies (3.2.20). □

3.3 Two estimates

As a preparation for the Hölder continuity results in Chapter 6, we derive some estimates for the linear SPDE (3.2.3). Due to technical reasons, we restrict to the case of $\rho(x,x) = \rho(0,0)$ and there exists a constant K such that

$$\int_U |\nabla h(y,x)|^2 \mu(dy) \leq K, \qquad \forall\, x \in \mathbb{R},$$

where

$$\rho(x,y) \equiv \int_U h(z,x)h(z,y)\mu(dz).$$

Denote the constant $\frac{1}{2}(1+\rho(0,0))$ by κ. In this case, the forward version of (3.2.3) becomes

$$\begin{aligned}u_t(x) = u_0(x) &+ \int_0^t \kappa \Delta u_s(x) ds \\ &+ \int_0^t \int_U \nabla u_s(x) h(y,x) W(dsdy),\end{aligned} \tag{3.3.1}$$

where W is a white noise random measure on $\mathbb{R}_+ \times U$ with intensity measure μ on U.

We study SPDE (3.3.1) where u_0 is either a real or a generalized function for different purposes. To this end, we need to introduce some notation taken from Krylov (1999). As we will consider the case of $p = 2$ (in his notation) only, the index p used by him is not needed here. For example, his H_p^n and $\|\cdot\|_{n,p}$ will become H_n and $\|\cdot\|_n$, respectively.

For $\alpha \in (0,1)$ and generalized function u on $\mathbb{R}$, let

$$(I-\Delta)^\alpha u = c(\alpha) \int_0^\infty \frac{e^{-t}T_t u - u}{t^{\alpha+1}} dt, \tag{3.3.2}$$

and

$$(I-\Delta)^{-\alpha}u = d(\alpha)\int_0^\infty t^{\alpha-1}e^{-t}T_t u dt, \tag{3.3.3}$$

where $c(\alpha)$ and $d(\alpha)$ are two constants and T_t is the Brownian semigroup. As being indicated by Krylov (1999), (3.3.2) and (3.3.3) are sufficient to define $(I-\Delta)^{n/2}$ consistently for any $n \in \mathbb{R}$ (cf. Krasnoselskii *et al* (1976)). In particular, for any $\alpha, \beta \in \mathbb{R}$,

$$(I-\Delta)^{\alpha}(I-\Delta)^{\beta} = (I-\Delta)^{\alpha+\beta}.$$

Let H_n be the spaces of Bessel potentials with norms

$$\|u\|_n \equiv \|(I-\Delta)^{n/2}u\|_0 \tag{3.3.4}$$

where $\|\cdot\|_0$ is the norm in L^2. Note that for n being a positive integer, $\|u\|_n$ is equivalent to the usual Sobolev norm on H_n defined in the last section.

The existence and uniqueness of the solution to (3.3.1) have been studied by Krylov (1999) in suitable Banach spaces. In the remaining of this section, we assume that this equation has a solution (the existence of which will also be evident from the applications in later chapters), and the aim of this section is to prove that, with the appropriate initial condition, the solution actually lies in the spaces which will be useful for our purposes.

Let $\beta \in [0,1)$ and $u_0 \in H_{\beta-1}$. For $f \in C_0^\infty(\mathbb{R})$, i.e., f is infinitely differentiable with compact support, we have

$$\begin{aligned}\langle u_r, f\rangle &= \langle u_0, f\rangle + \int_0^r \langle \kappa\Delta u_s, f\rangle\, ds \\ &\quad + \int_0^r\int_U \langle \nabla u_s h(y,\cdot), f\rangle\, W(dsdy)\end{aligned} \tag{3.3.5}$$

where $\langle u, f\rangle$ stands for the duality between Hilbert spaces H_{-n} and H_n.

Applying Itô's formula to $\langle u_r, f\rangle^2$, we arrive at

$$\begin{aligned}\langle u_r, f\rangle^2 &= \langle u_0, f\rangle^2 + \int_0^r 2\langle u_s, f\rangle \langle \kappa\Delta u_s, f\rangle\, ds \\ &\quad + \int_0^r\int_U 2\langle u_s, f\rangle\langle \nabla u_s h(y,\cdot), f\rangle\, W(dsdy) \\ &\quad + \int_0^r\int_U \langle \nabla u_s h(y,\cdot), f\rangle^2 \mu(dy)ds.\end{aligned} \tag{3.3.6}$$

Summing up f over a CONS of $H_{1-\beta}$, by (3.3.6) we get

$$\begin{aligned}\|u_r\|_{\beta-1}^2 &= \|u_0\|_{\beta-1}^2 + \int_0^r 2\langle u_s, \kappa\Delta u_s\rangle_{\beta-1}\, ds \\ &\quad + \int_0^r\int_U \|\nabla u_s h(y,\cdot)\|_{\beta-1}^2 \mu(dy)ds \\ &\quad + \int_0^r\int_U 2\langle u_s, \nabla u_s h(y,\cdot)\rangle_{\beta-1} W(dsdy).\end{aligned} \tag{3.3.7}$$

We first apply (3.3.7) for $\beta = 0$. The following lemmas will be used in the proof of Theorem 3.3.3. We abuse the notation a bit by denoting

$$\|h\|_{k,\infty}^2 \equiv \sum_{j=0}^{k} \sup_{x\in\mathbb{R}} \int_U |\nabla^k h(y,x)|^2 \mu(dy),$$

and say that $h \in \mathbb{H}_{k,\infty}$ if $\|h\|_{k,\infty} < \infty$.

Lemma 3.3.1. *If $h \in \mathbb{H}_{1,\infty}$, then for any $u \in H_0$,*

$$\int_U \|\nabla u h(y,\cdot)\|_{-1}^2 \mu(dy) \le \|h\|_1^2 \|u\|_0^2.$$

Proof. Note that

$$\begin{aligned}\int_U \|uh(y,\cdot)\|_0^2 \mu(dy) &= \int_U \int_{\mathbb{R}} u(x)^2 h(y,x)^2 dx \mu(dy)\\ &= \int_{\mathbb{R}} u(x)^2 \int_U h(y,x)^2 \mu(dy) dx\\ &\le \|h\|_{0,\infty}^2 \|u\|_0^2,\end{aligned}$$

and similarly,

$$\int_U \|u\nabla h(y,\cdot)\|_0^2 \mu(dy) \le \|\nabla h\|_{0,\infty}^2 \|u\|_0^2.$$

Thus,

$$\begin{aligned}&\int_U \|\nabla u h(y,\cdot)\|_{-1}^2 \mu(dy)\\ &= \int_U \sup_{\|f\|_1\le 1} \langle \nabla u, h(y,\cdot) f\rangle^2 \mu(dy)\\ &= \int_U \sup_{\|f\|_1\le 1} \langle u, -f'h(y,\cdot) - f\nabla h(y,\cdot)\rangle_0^2 \mu(dy)\\ &\le \int_U \sup_{\|f\|_1\le 1} \left(\|uh(y,\cdot)\|_0 \|f'\|_0 + \|u\nabla h(y,\cdot)\|_0 \|f\|_0\right)^2 \mu(dy)\\ &\le \int_U \left(\|uh(y,\cdot)\|_0^2 + \|u\nabla h(y,\cdot)\|_0^2\right) \mu(dy)\\ &= \|h\|_{1,\infty}^2 \|u\|_0^2,\end{aligned}$$

where $\nabla h(y,x)$ is the derivative in x, and the first inequality follows from the definition of the norm using duality. $\square$

To consider $u \in H_{-1}$, we need a stronger condition on h than that in the last lemma.

Lemma 3.3.2. *If $h \in \mathbb{H}_{2,\infty}$, then for any $u \in H_{-1}$,*

$$\int_U \langle u, \nabla u h(y,\cdot)\rangle_{-1}^2 \mu(dy) \leq K_1 \|u\|_{-1}^2.$$

Proof. Let $g = (I-\Delta)^{-1}u \in H_1$. By integration by parts, for $f \in H_2$, we have

$$\begin{aligned} &\langle u, f\nabla u\rangle_{-1} \\ &= \left\langle (I-\Delta)^{-1}u, f\nabla u\right\rangle_0 = \langle fg, \nabla u\rangle_0 \\ &= -\langle (fg)', g - g''\rangle_0 \\ &= -\langle f'g, g\rangle_0 - \langle fg', g\rangle_0 + \langle f'g, g''\rangle_0 + \langle fg', g''\rangle_0 \\ &= -\langle f'g, g\rangle_0 - \langle (f+f'')g, g'\rangle_0 + \langle fg', g''\rangle_0 - \langle f'g', g'\rangle_0 . \end{aligned} \tag{3.3.8}$$

Using (3.1.8), we get

$$\langle fg', g''\rangle_0 = -\frac{1}{2}\int_{\mathbb{R}} g'(x)^2 \nabla f(x) dx.$$

Replacing f by $h(y,\cdot)$ and taking integration with respect to $\mu(dy)$, we get

$$\begin{aligned} &\int_U \langle h(y,\cdot)g', g''\rangle_0^2 \mu(dy) \\ &= \frac{1}{4}\int_U \left|\int_{\mathbb{R}} g'(x)^2 \nabla h(y,x)dx\right|^2 \mu(dy) \\ &= \frac{1}{4}\int_U\int_{\mathbb{R}}\int_{\mathbb{R}} g'(x_1)^2 g'(x_2)^2 \nabla h(y,x_1)\nabla h(y,x_2) dx_1 dx_2 \mu(dy) \\ &\leq K_2 \|h\|_{1,\infty}^2 \int_{\mathbb{R}}\int_{\mathbb{R}} g'(x_1)^2 g'(x_2)^2 dx_1 dx_2 \\ &= K_2 \|h\|_{1,\infty}^2 \|g'\|_0^4. \end{aligned}$$

Applying Cauchy-Schwartz inequality to the other terms on RHS of (3.3.8), we have

$$\langle u, f\nabla u\rangle_{-1}^2 \leq K_3 \|f\|_{2,\infty}^2 \|g'\|_0^4. \tag{3.3.9}$$

Replacing f by $h(y,\cdot)$ and taking integration with respect to measure μ on both sides of (3.3.9), we get

$$\int_U \langle u, \nabla u h(y,\cdot)\rangle_1^2 \mu(dy) \leq K_4 \|g'\|_0^2, \tag{3.3.10}$$

where $K_1 > 0$ is a constant. As

$$\begin{aligned}\|g'\|_0 &= \|\nabla(I-\Delta)^{-1}u\|_0 \\ &\le K_5\|(I-\Delta)^{-\frac{1}{2}}u\|_0 \\ &= K_5\|u\|_{-1},\end{aligned}$$

the conclusion of the lemma follows from (3.3.10), where the inequality is achieved by the boundedness of operator $\nabla(I-\Delta)^{-\frac{1}{2}}$ (cf. Krylov (1999)).

□

Theorem 3.3.3. *Suppose that $p \ge 1$, $u_0 \in H_{-1}$, $h \in \mathbb{H}_{2,\infty}$ and $\|h\|^2_{1,\infty} < 2\kappa$. Then,*

$$\mathbb{E}\sup_{t\le T}\|u_t\|^{2p}_{-1} + \mathbb{E}\left(\int_0^T \|u_t\|_0^2 dt\right)^p \le K\|u_0\|^{2p}_{-1}. \tag{3.3.11}$$

Proof. Using a smoothing technique as in Section 3.1 if necessary, we may assume that $u_t \in H_0$. By (3.3.4) we have

$$\begin{aligned}\langle u, \kappa\Delta u\rangle_{-1} &= \kappa\langle u, \Delta u - u\rangle_{-1} + \kappa\|u\|^2_{-1} \\ &= -\kappa\|u\|_0^2 + \kappa\|u\|^2_{-1}.\end{aligned} \tag{3.3.12}$$

Using Lemma 3.3.1 we then have

$$\begin{aligned}&2\langle u, \kappa\Delta u\rangle_{-1} + \int_U \|\nabla u h(y,\cdot)\|^2_{-1}\mu(dy) \\ &\le -\left(2\kappa - \|h\|^2_{1,\infty}\right)\|u\|_0^2 + 2\kappa\|u\|^2_{-1}.\end{aligned} \tag{3.3.13}$$

It follows from Lemma 3.3.2 that

$$\begin{aligned}&\mathbb{E}\sup_{t\le r}\left|\int_0^t\int_U 2\langle u_s, \nabla u_s h(y,\cdot)\rangle_{-1} W(dsdy)\right|^p \\ &\le K\mathbb{E}\left(\int_0^r\int_U \langle u_s, \nabla u_s h(y,\cdot)\rangle^2_{-1}\mu(dy)ds\right)^{p/2} \\ &\le K\mathbb{E}\int_0^r \|u_s\|^{2p}_{-1}ds.\end{aligned} \tag{3.3.14}$$

The identity (3.3.7) with $\beta = 0$, together with (3.3.13) and (3.3.14), imply that

$$\begin{aligned}&\mathbb{E}\sup_{s\le r}\|u_s\|^{2p}_{-1} + \mathbb{E}\left(\int_0^r \|u_s\|_0^2 ds\right)^p \\ &\le K_1\mathbb{E}\|u_0\|^{2p}_{-1} + K_2\int_0^r \mathbb{E}\|u_s\|^{2p}_{-1}ds.\end{aligned} \tag{3.3.15}$$

Removing the second term on the LHS of (3.3.15), we get

$$\mathbb{E}\sup_{s\le r}\|u_s\|_{-1}^{2p}\le K_1\|u_0\|_{-1}^{2p}+K_2\int_0^r\mathbb{E}\|u_s\|_{-1}^{2p}ds. \tag{3.3.16}$$

It follows from Gronwall's inequality that

$$\mathbb{E}\sup_{s\le r}\|u_s\|_{-1}^{2p}\le K_1\|u_0\|_{-1}^{2p}e^{K_2r}.$$

Removing the first term on the LHS of (3.3.15), we get

$$\begin{aligned}\mathbb{E}\left(\int_0^r\|u_s\|_0^2ds\right)^p&\le K_1\|u_0\|_{-1}^{2p}+K_2\int_0^rK_1\|u_0\|_{-1}^{2p}e^{K_2r}ds\\&\le K_3\|u_0\|_{-1}^{2p}.\end{aligned} \tag{3.3.17}$$

Inequalities (3.3.16) and (3.3.17) then imply (3.3.11). □

In most applications discussed in this book, we take

$$u_0=\delta_y \text{ or } u_0=\delta_{y_1}-\delta_{y_2}.$$

It is well known that for any $y\in\mathbb{R}$ and $\alpha>\frac{1}{2}$, $\delta_y\in H_2^{-\alpha}$ (cf. Example 1 of Section 5.2 in the book of Barros-Neto (1973)). This justifies the applicability of the last theorem.

We will prove a stronger version of Theorem 3.3.3 which is useful in estimating time-increment of random field $X_t(y)$ to be studied in Chapter 5. To this end, we need the following two lemmas.

Lemma 3.3.4. *For $h\in\mathbb{H}_{1,\infty}$, $\alpha\in(0,\frac{1}{2}]$ and $\epsilon>0$, there exists a constant K such that for any $u\in H_{1-2\alpha}$,*

$$\int_U\|\nabla uh(y,\cdot)\|_{-2\alpha}^2\mu(dy)\le(1+\epsilon)\rho(0,0)\|\nabla u\|_{-2\alpha}^2+K\|u\|_0^2.$$

Proof. Note that

$$\begin{aligned}I(t,x,y)&\equiv T_t(u'h(y,\cdot))(x)-T_tu'(x)T_th(y,\cdot)(x)\\&=\frac{1}{2}\int_{\mathbb{R}}dz_1\int_{\mathbb{R}}dz_2(u'(z_1)-u'(z_2))\left(h(y,z_1)-h(y,z_2)\right)\\&\qquad\times p_t(x-z_1)p_t(x-z_2)\\&=\frac{1}{2}\int_{\mathbb{R}}dz_1\int_{\mathbb{R}}dz_2(u(z_1)-u(z_2))\left(\nabla h(y,z_1)-\nabla h(y,z_2)\right)\\&\qquad\times p_t(x-z_1)p_t(x-z_2)\\&\quad+\frac{1}{2}\int_{\mathbb{R}}dz_1\int_{\mathbb{R}}dz_2(u(z_1)-u(z_2))\left(h(y,z_1)-h(y,z_2)\right)\\&\qquad\times\frac{z_1-z_2}{t}p_t(x-z_1)p_t(x-z_2).\end{aligned}$$

Using the inequality $(a+b)^2 \le 2a^2 + 2b^2$, we see that there is a constant K_1,

$$\begin{aligned}
J &\equiv \int_{\mathbb{R}} dx \int_U \mu(dy)|I(t,x,y)|^2 \\
&\le K_1 \int_{\mathbb{R}} dx \int_U \mu(dy) \int_{\mathbb{R}} dz_1 \int_{\mathbb{R}} dz_2 (u(z_1) - u(z_2))^2 \\
&\qquad \times \left(\nabla h(y,z_1) - \nabla h(y,z_2)\right)^2 p_t(x-z_1)p_t(x-z_2) \\
&\quad + K_1 \int_{\mathbb{R}} dx \int_U \mu(dy) \int_{\mathbb{R}} dz_1 \int_{\mathbb{R}} dz_2 (u(z_1) - u(z_2))^2 \\
&\qquad \times \left| \frac{1}{z_1 - z_2} \int_{z_2}^{z_1} \nabla h(y,z) dz \right|^2 \frac{|z_1 - z_2|^2}{t} p_t(x-z_1)p_t(x-z_2).
\end{aligned}$$

Note that for any $\epsilon > 0$,

$$\begin{aligned}
\frac{x^2}{t} p_t(x) p_{(1+\epsilon)t}(x)^{-1} &= \frac{x^2}{t}\sqrt{1+\epsilon}\exp\left(\frac{x^2}{2(1+\epsilon)t} - \frac{x^2}{4t}\right) \\
&= \sqrt{1+\epsilon}\frac{x^2}{t}\exp\left(-\frac{\epsilon x^2}{4t(1+\epsilon)}\right) \\
&\le \sqrt{1+\epsilon}\sup_{y\ge 0}\left(y\exp\left(-\frac{\epsilon}{2(1+\epsilon)}y\right)\right).
\end{aligned}$$

Thus, we continue the estimate of J as follows:

$$\begin{aligned}
J &\le K_2 \int_{\mathbb{R}} dz_1 \int_{\mathbb{R}} dz_2 (u(z_1)^2 + u(z_2)^2) p_{2t}(z_1 - z_2) \\
&\quad + K_2 \int_{\mathbb{R}} dz_1 \int_{\mathbb{R}} dz_2 (u(z_1)^2 + u(z_2)^2) p_{(2+\epsilon)t}(z_1 - z_2) \\
&= K_3 \|u\|_0^2.
\end{aligned}$$

On the other hand,

$$\begin{aligned}
&\int_{\mathbb{R}} \int_U dx \mu(dy) \left| \int_0^\infty t^{\alpha-1} e^{-t} T_t u'(x) T_t h(y,\cdot)(x) dt \right|^2 \\
&= \int_{\mathbb{R}} \int_U dx \mu(dy) \int_0^\infty \int_0^\infty ds dt (ts)^{\alpha-1} e^{-(t+s)} T_s u'(x) T_t u'(x) \\
&\qquad \times T_t h(y,\cdot)(x) T_s h(y,\cdot)(x) \\
&\le \rho(0,0) \int_{\mathbb{R}} dx \int_0^\infty \int_0^\infty ds dt (ts)^{\alpha-1} e^{-(t+s)} T_s u'(x) T_t u'(x) \\
&= \rho(0,0) \|\nabla u\|_{-2\alpha}^2.
\end{aligned}$$

By the triangular inequality, we have

$$\begin{aligned}
&\left(\int_U \|\nabla uh(y,\cdot)\|^2_{-2\alpha}\mu(dy)\right)^{1/2}\\
&\le \left(\int_{\mathbb{R}}\int_U dx\mu(dy)\int_0^\infty t^{\alpha-1}e^{-t}I(t,x,y)^2dt\right)^{1/2} + \|\nabla u\|_{-2\alpha}\\
&\le K_4\|u\|_0 + \sqrt{\rho(0,0)}\|\nabla u\|_{-2\alpha}.
\end{aligned}$$

The conclusion then follows from the elementary inequality

$$(a+b)^2 \le (1+\epsilon)a^2 + \left(1+\epsilon^{-1}\right)b^2.$$

□

Lemma 3.3.5. *For $h \in \mathbb{H}_{1,\infty}$, there exists a constant K such that for any $0 \le u \in H_0$,*

$$\int_U \langle u, \nabla uh(y,\cdot)\rangle^2_{-2\alpha}\,\mu(dy) \le K\|u\|^2_{-2\alpha}\|u\|^2_0.$$

Proof. Note that

$$\begin{aligned}
&\int_U \langle u, \nabla uh(y,\cdot)\rangle^2_{-2\alpha}\,\mu(dy)\\
&= \int_U \left(\int_{\mathbb{R}} dx \int_0^\infty\int_0^\infty (ts)^{\alpha-1}e^{-(t+s)}dtdsT_tu(x)T_s(u'h(y,\cdot))(x)\right)^2\mu(dy)\\
&\le I_1 + I_2,
\end{aligned}$$

where

$$I_1 = \int_U \left(\int_{\mathbb{R}} dx \int_0^\infty\int_0^\infty (ts)^{\alpha-1}e^{-(t+s)}dtdsT_tu(x)T_su'(x)h(y,x)\right)^2\mu(dy)$$

and

$$\begin{aligned}
I_2 = \int_U \Bigg(\int_{\mathbb{R}} dx \int_0^\infty\int_0^\infty &(ts)^{\alpha-1}e^{-(t+s)}dtds\\
&\times T_tu(x)\left(T_s(u'h(y,\cdot))(x) - T_su'(x)h(y,x)\right)\Bigg)^2\mu(dy).
\end{aligned}$$

By integration by parts and changing the order of T_s and ∇, we get

$$\begin{aligned}
&\int_{\mathbb{R}} dx \int_0^\infty\int_0^\infty (ts)^{\alpha-1}e^{-(t+s)}dtdsT_tu(x)T_su'(x)h(y,x)\\
&= \frac{1}{2}\int_{\mathbb{R}} dx \int_0^\infty\int_0^\infty (ts)^{\alpha-1}e^{-(t+s)}dtdsT_tu(x)\nabla h(y,x)T_su(x).
\end{aligned}$$

Thus,

$$\begin{aligned}
I_1 &\le K_1 \int_{\mathbb{R}} dx \int_0^\infty \int_0^\infty (ts)^{\alpha-1} e^{-(t+s)} dt ds T_t u(x) T_s u(x) \\
&\quad \times \int_{\mathbb{R}} dx' \int_0^\infty \int_0^\infty (t's')^{\alpha-1} e^{-(t'+s')} dt' ds' T_{t'} u(x') T_{s'} u(x') \\
&= K_1 \|u\|_{-2\alpha}^4 \\
&\le K_2 \|u\|_{-2\alpha}^2 \|u\|_0^2.
\end{aligned}$$

Note that we used the non-negativity of $u(x)$ in the first inequality above.

Now we estimate I_2. Note that

$$\begin{aligned}
I_2 = \int_{\mathbb{R}} dx \int_{\mathbb{R}} dx' \int_{\mathbb{R}_+} dt \int_{\mathbb{R}_+} ds \int_{\mathbb{R}_+} dt' \int_{\mathbb{R}_+} ds' (tst's')^{\alpha-1} \\
\times e^{-(t+s+t'+s')} T_t u(x) T_{t'} u(x') J(x,x'),
\end{aligned}$$

where

$$\begin{aligned}
J(x,x') &= \int_U \lambda(dy) \int_{\mathbb{R}^2} dz dz' p_s(x-z)(h(y,z)-h(y,x)) u'(z) \\
&\qquad \times p_{s'}(x'-z')(h(y,z')-h(y,x')) u'(z') \\
&= \int_{\mathbb{R}^2} dz dz' p_s(x-z) p_{s'}(x'-z') u'(z) u'(z') \\
&\qquad \times \left(\rho(z,z') - \rho(x,z') - \rho(z,x') + \rho(x,x')\right).
\end{aligned}$$

By integration by part, we can continue the calculation with

$$J(x,x') = J_1 + J_2 + J_3 + J_4,$$

where

$$\begin{aligned}
J_1 = \int_{\mathbb{R}^2} dz dz' \nabla_z p_s(x-z) \nabla_{z'} p_{s'}(x'-z') u(z) u(z') \\
\times \left(\rho(z,z') - \rho(x,z') - \rho(z,x') + \rho(x,x')\right),
\end{aligned}$$

$$J_2 = \int_{\mathbb{R}^2} dz dz' \nabla_z p_s(x-z) p_{s'}(x'-z') \left(\nabla_{z'} \rho(x,z') - \nabla_{z'} \rho(z,z')\right) u(z) u(z'),$$

$$J_3 = \int_{\mathbb{R}^2} dz dz' p_s(x-z) \nabla_{z'} p_{s'}(x'-z') \left(\nabla_z \rho(z,z') - \nabla_z \rho(z,x')\right) u(z) u(z'),$$

and

$$J_4 = -\int_{\mathbb{R}^2} dz dz' p_s(x-z) p_{s'}(x'-z') \nabla_z \nabla_{z'} \rho(z,z') u(z) u(z').$$

By Lipschitz property, we see that $\nabla_z\rho$ is bounded, and hence

$$J_2 \leq K_3 \int_{\mathbb{R}^2} dzdz' p_s(x-z)\frac{|x-z|^2}{s} p_{s'}(x'-z')u(z)u(z').$$

Thus, for some $\epsilon > 0$ we have

$$\begin{aligned} J_2 &\leq K_4 \int_{\mathbb{R}^2} dzdz' p_{(1+\epsilon)s}(x-z)p_{s'}(x'-z')u(z)u(z') \\ &= K_4 T_{(1+\epsilon)s}u(x)T_{s'}u(x'). \end{aligned}$$

Let $\mathcal{G}u(x) = \int_0^\infty s^{\alpha-1}e^{-s}T_{(1+\epsilon)s}u(x)ds$. Then the corresponding term of J_2 in I_2 is bounded, up to a constant multiplication, by

$$\begin{aligned} &\|u\|^2_{-2\alpha}\int_{\mathbb{R}} dx \int_0^\infty \int_0^\infty (ts)^{\alpha-1}e^{-(t+s)}dtdsT_tu(x)T_{(1+\epsilon)s}u(x) \\ &= \|u\|^2_{-2\alpha}\left\langle (I-\Delta)^{-\alpha}u, \mathcal{G}u\right\rangle_0 \\ &\leq \|u\|^3_{-2\alpha}\|\mathcal{G}u\|_0 \\ &\leq \|u\|^3_{-2\alpha}\left(\int_0^\infty \int_0^\infty (ts)^{\alpha-1}e^{-(t+s)}dtds\|u\|_0^2\right)^{1/2} \\ &\leq K_5\|u\|^3_{-2\alpha}\|u\|_0 \\ &\leq K_6\|u\|^2_{-2\alpha}\|u\|^2_0. \end{aligned}$$

The terms J_i for $j = 1, 3, 4$ can be estimated similarly. □

Now, we are ready to derive the estimate which will be useful in estimating the time-increment of the density random field which we will study in Chapter 5.

Theorem 3.3.6. *Suppose that the conditions of Theorem 3.3.3 hold. Then for $\beta \in [0, 1/2], u_0 = \delta_z$ and $p \geq 1$ we have*

$$\mathbb{E}\sup_{t\leq T}\|u_t\|^{2p}_{\beta-1} + \mathbb{E}\left(\int_0^T \|u_t\|^2_\beta dt\right)^p \leq K\|\delta_z\|^{2p}_{\beta-1}. \tag{3.3.18}$$

Proof. Similar to Theorem 3.3.3 we may assume that $u_t \in H_\beta$ a.s.. Further, using a stopping argument if necessary we may and will assume that the LHS of (3.3.18) is finite. Denoting $1-\beta = 2\alpha$ for simplicity, by (3.3.7) and Lemma 3.3.4 we obtain,

$$\begin{aligned}\|u_r\|_{-2\alpha}^2 \le& \|u_0\|_{-2\alpha}^2 - (\epsilon\rho(0,0)-1)\int_0^r \|\nabla u_s\|_{-2\alpha}^2 ds + K_1\int_0^r \|u_s\|_0^2 ds\\ &+\int_0^r\int_U 2\langle u_s, \nabla u_s h(y,\cdot)\rangle_{\beta-1} W(dsdy)\\ \le& \|u_0\|_{-2\alpha}^2 - (\epsilon\rho(0,0)-1)\int_0^r \|u_s\|_{1-2\alpha}^2 ds + K_2\int_0^r \|u_s\|_0^2 ds\\ &+\int_0^r\int_U 2\langle u_s, \nabla u_s h(y,\cdot)\rangle_{\beta-1} W(dsdy),\end{aligned}$$

where K_1 and K_2 are constants, and the last inequality follows from the following calculation for any $u \in H_{1-2\alpha}$,

$$\begin{aligned}\|\nabla u\|_{-2\alpha}^2 &= d(\alpha)^2\left\langle \int_0^\infty t^{\alpha-1}e^{-t}T_t u' dt, \int_0^\infty t^{\alpha-1}e^{-t}T_t u' dt\right\rangle_0\\ &= -d(\alpha)^2\left\langle \int_0^\infty t^{\alpha-1}e^{-t}T_t u dt, \Delta\int_0^\infty t^{\alpha-1}e^{-t}T_t u dt\right\rangle_0\\ &= d(\alpha)^2\left\langle \int_0^\infty t^{\alpha-1}e^{-t}T_t u dt, (I-\Delta)\int_0^\infty t^{\alpha-1}e^{-t}T_t u dt\right\rangle_0\\ &\quad - d(\alpha)^2\left\langle \int_0^\infty t^{\alpha-1}e^{-t}T_t u dt, \int_0^\infty t^{\alpha-1}e^{-t}T_t u dt\right\rangle_0\\ &= \|(I-\Delta)^{\frac12}(I-\Delta)^{-\alpha}u\|_2^2 - \|(I-\Delta)^{-\alpha}u\|_2^2\\ &= \|u\|_{1-2\alpha}^2 - \|u\|_{-2\alpha}^2\\ &\ge \|u\|_{1-2\alpha}^2 - \|u\|_0^2.\end{aligned}$$

Let ϵ be small enough such that $\epsilon\rho(0,0)-1<0$. Then, there exist constants K_3 and K_4 such that

$$\begin{aligned}&\mathbb{E}\sup_{t\le r}\|u_t\|_{\beta-1}^{2p} + \mathbb{E}\left(\int_0^r \|u_t\|_\beta^2 dt\right)^p\\ &\le K_3\|u_0\|_{-2\alpha}^{2p} + K_3\mathbb{E}\left(\int_0^r \|u_s\|_0^2 ds\right)^p\\ &\quad +K_3\mathbb{E}\left(\int_0^r\int_U 2\langle u_s, \nabla u_s h(y,\cdot)\rangle_{\beta-1}^2 \mu(dy)ds\right)^{p/2}\\ &\le K_4\|u_0\|_{-2\alpha}^{2p} + K_4\mathbb{E}\left(\int_0^r\int_U 2\langle u_s, \nabla u_s h(y,\cdot)\rangle_{\beta-1}^2 \mu(dy)ds\right)^{p/2},\end{aligned}$$

where the last inequality follows from Theorem 3.3.3 and the fact that $\|u_0\|_{-1} \le \|u_0\|_{-2\alpha}$.

By Lemma 3.3.5, we get

$$
\begin{aligned}
&\mathbb{E}\sup_{s\le r}\|u_s\|_{\beta-1}^{2p}+\mathbb{E}\left(\int_0^r\|u_s\|_\beta^2 ds\right)^p\\
\le\ & K_4\|\delta_z\|_{\beta-1}^{2p}+K_5\mathbb{E}\left(\int_0^r\|u_s\|_{\beta-1}^2\|u_s\|_0^2 ds\right)^{p/2}\\
\le\ & K_4\|\delta_z\|_{\beta-1}^{2p}+K_5\mathbb{E}\left(\sup_{s\le r}\|u_s\|_{\beta-1}^2\int_0^r\|u_s\|_0^2 ds\right)^{p/2}\\
\le\ & K_4\|\delta_z\|_{\beta-1}^{2p}+\frac{1}{2}\mathbb{E}\sup_{s\le r}\|u_s\|_{\beta-1}^{2p}+K_6\mathbb{E}\left(\int_0^r\|u_s\|_0^2 ds\right)^p\\
\le\ & K_4\|\delta_z\|_{\beta-1}^{2p}+\frac{1}{2}\mathbb{E}\sup_{s\le r}\|u_s\|_{\beta-1}^{2p}+K_7\|\delta_z\|_{-1}^{2p}.
\end{aligned}
$$

The conclusion then follows from easy calculations. □

3.4 Historical remarks

The material of Section 3.1 is taken from Kurtz and Xiong (1999). A similar idea was employed by Kotelenez (1995). The material of Section 3.3 is based on the results of Li, Wang, Xiong and Zhou (2012). As we already mentioned in this chapter, some of the notations are taken from Krylov (1999).

Chapter 4

Particle Representations for a Class of Nonlinear SPDEs

In this chapter, we consider the particle representation for a class of nonlinear SPDEs. Besides its usefulness in deriving numerical schemes for the solutions of such SPDEs, it also provides an efficient way of proving uniqueness for nonlinear SPDEs. More specifically, the uniqueness problem of a nonlinear SPDE can be decomposed into two easier ones: that of a corresponding linear SPDE and that of a system of stochastic differential equations.

4.1 Introduction

We consider a class of nonlinear stochastic partial differential equations of the form

$$dv_t(x) = \left(\frac{1}{2}\Delta\left[a(x,v_t)v_t(x)\right] - \nabla\left[b(x,v_t)v_t(x)\right] + d(x,v_t)v_t(x)\right)dt \\ - \int_U \left(\beta(x,v_t,u)v_t(x) + \nabla\left[\alpha(x,v(t,\cdot),u)\right]\right) W(dtdu), \tag{4.1.1}$$

where W is a white noise random measure on $[0,\infty)\times U$. We are interested in the representations of solutions in terms of weighted empirical measures of the form

$$V_t = \lim_{n\to\infty}\frac{1}{n}\sum_{i=1}^n A_t^i\delta_{X_t^i} \tag{4.1.2}$$

where δ_x is the Dirac measure at x and the limit exists in the weak* topology on $\mathcal{M}_G(\mathbb{R})$, the collection of all finite signed Borel measures on $\mathbb{R}$. We think of $\{X_t^i : t\geq 0, i\in\mathbb{N}\}$ as a system of particles with locations in $\mathbb{R}$ and time-varying weights $\{A_t^i : t\geq 0, i\in\mathbb{N}\}$.

Suppose $\{X^i, A^i, V\}$ is governed by the following equations with V_s given by (4.1.2):

$$X_t^i = X_0^i + \int_0^t \sigma(X_s^i, V_s)dB_i(s) + \int_0^t c(X_s^i, V_s)ds + \int_0^t \int_U \alpha(X_s^i, V_s, u)W(dsdu) \tag{4.1.3}$$

and

$$A_t^i = A_0^i + \int_0^t A_s^i \gamma(X_s^i, V_s)dB_i(s) + \int_0^t A_s^i d(X_s^i, V_s)ds + \int_0^t \int_U A_s^i \beta(X_s^i, V_s, u)W(dsdu) \tag{4.1.4}$$

where c, d, σ, γ : $\mathbb{R} \times \mathcal{M}_G(\mathbb{R}) \to \mathbb{R}$ and α, β : $\mathbb{R} \times \mathcal{M}_G(\mathbb{R}) \times U \to \mathbb{R}$ are measurable mappings, B_i's are independent, standard $\mathbb{R}$-valued Brownian motions and W, independent of $\{B_i\}$, is a white noise random measure with

$$\mathbb{E}[W([0,t] \times A)W([0,t] \times B)] = \mu(A \cap B)t,$$

and μ is a σ-finite measure on the measurable space $(U, \mathcal{U})$.

Assume that $\{(A_0^i, X_0^i)\}_{i=1}^\infty$ is exchangeable (for example, i.i.d.) and independent of $\{B_i\}$ and W. Applying Itô's formula to (4.1.3) and (4.1.4), for every $\phi \in C_b^2(\mathbb{R})$, we have

$$\begin{aligned} A_t^i \phi(X_t^i) = A_0^i \phi(X_0^i) &+ \int_0^t A_s^i \phi(X_s^i)\gamma(X_s^i, V_s)dB_i(s) \\ &+ \int_0^t A_s^i \phi(X_s^i) d(X_s^i, V_s)ds \\ &+ \int_0^t \int_U A_s^i \phi(X_s^i)\beta(X_s^i, V_s, u)W(dsdu) \\ &+ \int_0^t A_s^i L_{V_s}\phi(X_s^i)ds \\ &+ \int_0^t A_s^i \phi'(X_s^i)\sigma(X_s^i, V_s)dB_i(s) \\ &+ \int_0^t \int_U A_s^i \phi'(X_s^i)\alpha(X_s^i, V_s, u)W(dsdu), \end{aligned} \tag{4.1.5}$$

where

$$L_v\phi(x) = \frac{1}{2}a(x,v)\phi''(x) + b(x,v)\phi'(x)$$

with

$$a(x,v) = \sigma(x,v)\sigma(x,v) + \int_U \alpha(x,v,u)\alpha(x,v,u)\mu(du)$$

and

$$b(x,v) = c(x,v) + \sigma(x,v)\gamma(x,v) + \int_U \beta(x,v,u)\alpha(x,v,u)\mu(du).$$

Averaging both sides of (4.1.5) and taking $n \to \infty$, we see that V given by (4.1.2) satisfies: $\forall \phi \in C_b^2(\mathbb{R})$,

$$\langle V_t, \phi\rangle = \langle V_0, \phi\rangle + \int_0^t \langle V_s, d(\cdot, V_s)\phi + L_{V_s}\phi, V_s\rangle\, ds \tag{4.1.6}$$

$$+ \int_0^t \int_U \langle \beta(\cdot, V_s, u)\phi + \alpha(\cdot, V_s, u)\phi', V_s\rangle\, W(dsdu),$$

and, hence, is a weak solution of SPDE (4.1.1) where, if it exists, v is the density

$$V_t(B) = \int_B v_t(x)dx, \qquad \forall\ B \in \mathcal{B}(\mathbb{R}).$$

Our goal is to give conditions under which there exists a unique solution of system (4.1.3, 4.1.4) and as a consequence obtain existence and uniqueness of the solution to SPDE (4.1.6).

This chapter is organized as follows: In Section 4.2, we prove that the system (4.1.2)-(4.1.4) has a unique solution. Since the coefficients $a\gamma(x,v)$, $ad(x,v)$ and $a\beta(x,v,u)$ are not Lipschitz in (a,x,v), the system does not satisfy a global Lipschitz condition, we cannot directly apply the results developed by Kurtz and Protter (1996). Instead, a truncation technique is employed (cf. Theorem 4.2.2). In Section 4.3, we prove existence and uniqueness for the solution to SPDE (4.1.6) and achieve this goal by considering a corresponding linear equation of the form (3.1.1) first. As a by-product from this linear equation, the existence of density $v_t(x)$ is obtained. Moreover, the uniqueness of the solution to system (4.1.2)-(4.1.4) and that for linear equation, implies the uniqueness of the solution to nonlinear SPDE (4.1.1).

4.2 Solution for the system

In this section, we establish existence and uniqueness for the solution of the system (4.1.2)-(4.1.4). For $\nu_1, \nu_2 \in \mathcal{M}_F(\mathbb{R})$, the Wasserstein metric is defined by

$$\rho(\nu_1, \nu_2) = \sup\{|\langle \phi, \nu_1\rangle - \langle \phi, \nu_2\rangle| : \phi \in \mathbb{B}_1\}$$

where

$$\mathbb{B}_1 = \{\phi : |\phi(x) - \phi(y)| \le |x - y|, |\phi(x)| \le 1, \forall x, y \in \mathbb{R}\}.$$

Note that metric ρ determines the topology of weak convergence and convergence of first moments on $\mathcal{M}_F(\mathbb{R})$.

We assume that the coefficients satisfy the following conditions (S1) and (S2):

(S1) There exists a constant K such that for each $x \in \mathbb{R}$, $\nu \in \mathcal{M}_G(\mathbb{R})$

$$\begin{aligned}&|\sigma(x,\nu)|^2 + |c(x,\nu)|^2 + \int_U |\alpha(x,\nu,u)|^2 \mu(du)\\&+ |\gamma(x,\nu)|^2 + |d(x,\nu)|^2 + \int_U \beta(x,\nu,u)^2 \mu(du) \le K^2.\end{aligned}$$

(S2) For each $x_1, x_2 \in \mathbb{R}$, $\nu_1, \nu_2 \in \mathcal{M}_G(\mathbb{R})$ and any representation

$$\nu_i = \nu_i^+ - \nu_i^-, \qquad \nu_i^+, \nu_i^- \in \mathcal{M}_F(\mathbb{R}),$$

we have

$$\begin{aligned}&|\sigma(x_1,\nu_1) - \sigma(x_2,\nu_2)|^2 + |c(x_1,\nu_1) - c(x_2,\nu_2)|^2\\&+ |\gamma(x_1,\nu_1) - \gamma(x_1,\nu_1)|^2 + \int_U |\alpha(x_1,\nu_1,u) - \alpha(x_2,\nu_2,u)|^2 \mu(du)\\&+ |d(x_1,\nu_1) - d(x_2,\nu_2)|^2 + \int_U |\beta(x_1,\nu_1,u) - \beta(x_2,\nu_2,u)|^2 \mu(du)\\&\le K^2(|x_1 - x_2|^2 + \rho(\nu_1^+,\nu_2^+)^2 + \rho(\nu_1^-,\nu_2^-)^2).\end{aligned}$$

Let (X, A, V) be a solution of (4.1.2)-(4.1.4). In order to apply the Lipschitz condition (S2), we identify a canonical decomposition

$$V_t = V_t^+ - V_t^-.$$

For the simplicity of notation, define

$$M_t = \int_0^t \gamma(X_s^i, V_s) dB_i(s) + \int_0^t \int_U \beta(X_s^i, V_s, u) W(dsdu).$$

Then M_t is a martingale with quadratic variation process

$$\begin{aligned}\langle M \rangle_t &= \int_0^t |\gamma(X_s^i, V_s)|^2 ds + \int_0^t \int_U \beta(X_s^i, V_s, u)^2 \mu(du) ds\\&\le K^2 t.\end{aligned}$$

An application of Itô's formula shows that the solution of (4.1.4) is given by

$$A_t^i = A_0^i \exp\left(M_t - \frac{1}{2} \langle M \rangle_t + \int_0^t d(X_s^i, V_s) ds \right). \tag{4.2.1}$$

Note that if $A_0^i > 0$, then $A_t^i > 0$ for all $t > 0$ and similarly if $A_0^i < 0$, then $A_t^i < 0$ for all $t > 0$. Let

$$A_t^{i,+} = \begin{cases} A_t^i & \text{if } A_t^i > 0, \\ 0, & \text{otherwise,} \end{cases}$$

and let

$$A_t^{i,-} = \begin{cases} -A_t^i & \text{if } A_t^i < 0, \\ 0, & \text{otherwise.} \end{cases}$$

Then we define

$$V_t^+ = \lim_{n\to\infty} \frac{1}{n} \sum_{i=1}^n A_t^{i,+} \delta_{X_t^i} \qquad \text{and} \qquad V_t^- = \lim_{n\to\infty} \frac{1}{n} \sum_{i=1}^n A_t^{i,-} \delta_{X_t^i}. \tag{4.2.2}$$

A truncation argument will require the following estimate.

Proposition 4.2.1. *Suppose that Assumption (S1) holds and*

$$\mathbb{E}|A_0^1|^2 + \mathbb{E}|X_0^1|^2 < \infty. \tag{4.2.3}$$

If (X, A, V) is a solution of (4.1.2)-(4.1.4), then for every $t \geq 0$,

$$\mathbb{E} \sup_{0\leq s\leq t} \left(|A_s^i|^2 + |X_s^i|^2\right) < \infty, \tag{4.2.4}$$

where the LHS does not depend on i.

Proof. By Doob's inequality, we have

$$\begin{aligned}
&\mathbb{E} \sup_{0\leq s\leq t} |X_s^i|^2 \\
&\leq 4\mathbb{E}|X_0^i|^2 + 16\mathbb{E}\int_0^t |\sigma(X_s^i, V_s)|^2 ds \\
&\quad + 4t\mathbb{E}\int_0^t |c(X_s^i, V_s)|^2 ds + 16\mathbb{E}\int_0^t \int_U |\alpha(X_s^i, V_s, u)|^2 \mu(du) ds \\
&\leq 4\mathbb{E}|X_0^i|^2 + 32K^2 t + 4K^2 t^2 < \infty.
\end{aligned}$$

By (4.2.1), we have

$$|A_t^i|^2 = |A_0^i|^2 \exp\left(2M_t - \langle M\rangle_t + \int_0^t 2d(X_s^i, V_s) ds\right).$$

Keeping in mind that $\exp(2M_t - 2\langle M\rangle_t)$ is a martingale and using the bounds on $\langle M\rangle$ and d, and Doob's inequality, we have

$$\mathbb{E} \sup_{0\leq s\leq t} |A_s^i|^2 \leq 4e^{2Kt+K^2 t}\mathbb{E}|A_0^i|^2.$$

The independency of i follows from the exchangeability of the system. $\square$

Theorem 4.2.2. *Under Assumptions (S1), (S2) and (4.2.3), system (4.1.2)-(4.1.4) has at most one solution.*

Proof. Let (X, A, V) and $(\tilde{X}, \tilde{A}, \tilde{V})$ be two solutions of (4.1.2)-(4.1.4) with the same initial conditions, and define V^+, V^-, $\tilde{V}^+$, and $\tilde{V}^-$ as in (4.2.2).

Recall that by the exchangeability, we have

$$\lim_{n\to\infty} \frac{1}{n}\sum_{i=1}^{n} |A_t^i|^2 \qquad \text{exists.}$$

Let

$$\tau_m = \inf\left\{t : \lim_{n\to\infty} \frac{1}{n}\sum_{i=1}^{n} |A_t^i|^2 > m^2\right\}.$$

The stopping time $\tilde{\tau}_m$ is defined similarly. Let $\eta_m = \tau_m \wedge \tilde{\tau}_m$. Then

$$\begin{aligned}
&\mathbb{E}|X^i_{t\wedge\eta_m} - \tilde{X}^i_{t\wedge\eta_m}|^2 \qquad (4.2.5)\\
&\le 12\mathbb{E}\int_0^t |\sigma(X_s^i, V_s) - \sigma(\tilde{X}_s^i, \tilde{V}_s)|^2 1_{s\le\eta_m} ds\\
&\quad + 3t\mathbb{E}\int_0^t |c(X_s^i, V_s) - c(\tilde{X}_s^i, \tilde{V}_s)|^2 1_{s\le\eta_m} ds\\
&\quad + 12\mathbb{E}\int_0^t\int_U |\alpha(X_s^i, V_s, u) - \alpha(\tilde{X}_s^i, \tilde{V}_s, u)|^2 1_{s\le\eta_m}\mu(du)ds\\
&\le 3K^2(8+t)\mathbb{E}\int_0^t \left(|X_s^i - \tilde{X}_s^i|^2 + \rho(V_s^+, \tilde{V}_s^+)^2 + \rho(V_s^-, \tilde{V}_s^-)^2\right) 1_{s\le\eta_m} ds.
\end{aligned}$$

For $s \le \eta_m$, we estimate

$$\begin{aligned}
\rho(V_s^+, \tilde{V}_s^+) &= \sup_{\phi\in\mathbb{B}_1}\left|\lim_{n\to\infty}\frac{1}{n}\sum_{i=1}^{n}(A_s^{i,+}\phi(X_s^i) - \tilde{A}_s^{i,+}\phi(\tilde{X}_s^i))\right|\\
&\le \sup_{\phi\in\mathbb{B}_1}\lim_{n\to\infty}\frac{1}{n}\sum_{i=1}^{n} A_s^{i,+}|\phi(X_s^i) - \phi(\tilde{X}_s^i)|\\
&\quad + \sup_{\phi\in\mathbb{B}_1}\lim_{n\to\infty}\frac{1}{n}\sum_{i=1}^{n}|A_s^{i,+} - \tilde{A}_s^{i,+}|\left|\phi(\tilde{X}_s^i)\right|\\
&\le \lim_{n\to\infty}\frac{1}{n}\sum_{i=1}^{n} A_s^{i,+}|X_s^i - \tilde{X}_s^i|\\
&\quad + \lim_{n\to\infty}\frac{1}{n}\sum_{i=1}^{n}|A_s^{i,+} - \tilde{A}_s^{i,+}|
\end{aligned}$$

and a similar estimate holds for $\rho(V_t^-, \tilde{V}_t^-)$. Consequently,

$$\begin{aligned}
&\rho(V_s^+, \tilde{V}_s^+) + \rho(V_s^-, \tilde{V}_s^-) \\
&\leq \left(\lim_{n\to\infty} \frac{1}{n}\sum_{i=1}^n |A_s^i|^2\right)^{\frac{1}{2}} \left(\lim_{n\to\infty} \frac{1}{n}\sum_{i=1}^n |X_s^i - \tilde{X}_s^i|^2\right)^{\frac{1}{2}} \\
&\quad + \lim_{n\to\infty} \frac{1}{n}\sum_{i=1}^n |A_s^i - \tilde{A}_s^i| \\
&\leq m\left(\lim_{n\to\infty} \frac{1}{n}\sum_{i=1}^n |X_s^i - \tilde{X}_s^i|^2\right)^{\frac{1}{2}} + \lim_{n\to\infty} \frac{1}{n}\sum_{i=1}^n |A_s^i - \tilde{A}_s^i|.
\end{aligned}$$

Now let

$$f_m(t) = \mathbb{E}|X_{t\wedge\eta_m}^i - \tilde{X}_{t\wedge\eta_m}^i|^2,$$

and

$$g_m(t) = \mathbb{E}\left[\left(\lim_{n\to\infty} \frac{1}{n}\sum_{i=1}^n |A_{t\wedge\eta_m}^i - \tilde{A}_{t\wedge\eta_m}^i|\right)^2\right].$$

By (4.2.5) and Fatou's lemma, we have,

$$f_m(t) \leq 3K^2(8+t)\int_0^t \left(f_m(s) + 2m^2 f_m(s) + 2g_m(s)\right) ds. \tag{4.2.6}$$

By (4.2.1) and making use of the fact that

$$|e^x - e^y| \leq (e^x \vee e^y)|x-y|,$$

we have

$$\begin{aligned}
&|A_t^i - \tilde{A}_t^i| \\
&= (|A_t^i| \vee |\tilde{A}_t^i|)\left|\int_0^t (\gamma(X_s^i, V_s) - \gamma(\tilde{X}_s^i, \tilde{V}_s))dB_i(s)\right. \\
&\quad + \int_0^t\int_U (\beta(X_s^i, V_s, u) - \beta(\tilde{X}_s^i, \tilde{V}_s, u))W(dsdu) \\
&\quad + \int_0^t (d(X_s^i, V_s) - d(\tilde{X}_s^i, \tilde{V}_s))ds \\
&\quad - \frac{1}{2}\int_0^t (|\gamma(X_s^i, V_s)|^2 - |\gamma(\tilde{X}_s^i, \tilde{V}_s)|^2)ds \\
&\quad \left. - \frac{1}{2}\int_0^t\int_U (\beta(X_s^i, V_s, u)^2 - \beta(\tilde{X}_s^i, \tilde{V}_s, u))^2\mu(du)ds\right|.
\end{aligned}$$

Hence, for $t \le \eta_m$

$$
\begin{aligned}
&\left(\lim_{n\to\infty}\frac{1}{n}\sum_{i=1}^{n}|A_t^i-\tilde{A}_t^i|\right)^2\\
\le&\lim_{n\to\infty}\frac{1}{n}\sum_{i=1}^{n}\left(|A_t^i|^2\vee|\tilde{A}_t^i|^2\right)\frac{1}{n}\sum_{i=1}^{n}\left|\int_0^t(\gamma(X_s^i,V_s)-\gamma(\tilde{X}_s^i,\tilde{V}_s))dB_i(s)\right.\\
&+\int_0^t\int_U(\beta(X_s^i,V_s,u)-\beta(\tilde{X}_s^i,\tilde{V}_s,u))W(dsdu)\\
&+\int_0^t(d(X_s^i,V_s)-d(\tilde{X}_s^i,\tilde{V}_s))ds\\
&-\frac{1}{2}\int_0^t(|\gamma(X_s^i,V_s)|^2-|\gamma(\tilde{X}_s^i,\tilde{V}_s)|^2)ds\\
&\left.-\frac{1}{2}\int_0^t\int_U(\beta(X_s^i,V_s,u)^2-\beta(\tilde{X}_s^i,\tilde{V}_s,u)^2)\mu(du)ds\right|^2.
\end{aligned}
$$

Using the inequality

$$(a_1+\cdots+a_5)^2\le 5(a_1^2+\cdots+a_5^2),$$

we may continue the estimate with

$$\left(\lim_{n\to\infty}\frac{1}{n}\sum_{i=1}^{n}|A_t^i-\tilde{A}_t^i|\right)^2\le 10m^2\left(I_t^1+\cdots+I_t^5\right), \tag{4.2.7}$$

where

$$I_t^1=\lim_{n\to\infty}\frac{1}{n}\sum_{i=1}^{n}\left|\int_0^t(\gamma(X_s^i,V_s)-\gamma(\tilde{X}_s^i,\tilde{V}_s))dB_i(s)\right|^2,$$

$$I_t^2=\lim_{n\to\infty}\frac{1}{n}\sum_{i=1}^{n}\left|\int_0^t\int_U(\beta(X_s^i,V_s,u)-\beta(\tilde{X}_s^i,\tilde{V}_s,u))W(dsdu)\right|^2,$$

$$I_t^3=\lim_{n\to\infty}\frac{1}{n}\sum_{i=1}^{n}\left|\int_0^t(d(X_s^i,V_s)-d(\tilde{X}_s^i,\tilde{V}_s))ds\right|^2,$$

$$I_t^4=\lim_{n\to\infty}\frac{1}{n}\sum_{i=1}^{n}\left|\frac{1}{2}\int_0^t(|\gamma(X_s^i,V_s)|^2-|\gamma(\tilde{X}_s^i,\tilde{V}_s)|^2)ds\right|^2,$$

and

$$I_t^5=\lim_{n\to\infty}\frac{1}{n}\sum_{i=1}^{n}\left|\frac{1}{2}\int_0^t\int_U(\beta(X_s^i,V_s,u)^2-\beta(\tilde{X}_s^i,\tilde{V}_s,u)^2)\mu(du)ds\right|^2.$$

Note that

$$I_t^3 \le t\int_0^t (d(X_s^i, V_s) - d(\tilde{X}_s^i, \tilde{V}_s))^2 ds$$
$$\le K^2 t\int_0^t \left(|X_s^i - \tilde{X}_s^i|^2 + \rho(V_s^+, \tilde{V}_s^+)^2 + \rho(V_s^-, \tilde{V}_s^-)^2\right) ds,$$

$$I_t^4 \le \frac{1}{4}t\int_0^t (|\gamma(X_s^i, V_s)|^2 - |\gamma(\tilde{X}_s^i, \tilde{V}_s)|^2)^2 ds$$
$$\le \frac{1}{4}t\int_0^t (2K)^2|\gamma(X_s^i, V_s) - \gamma(\tilde{X}_s^i, \tilde{V}_s)|^2 ds$$
$$\le K^4 t\int_0^t \left(|X_s^i - \tilde{X}_s^i|^2 + \rho(V_s^+, \tilde{V}_s^+)^2 + \rho(V_s^-, \tilde{V}_s^-)^2\right) ds,$$

and

$$I_t^5 \le \frac{1}{4}t\int_0^t \left(\int_U (\beta(X_s^i, V_s, u)^2 - \beta(\tilde{X}_s^i, \tilde{V}_s, u)^2)\mu(du)\right)^2 ds$$
$$\le \frac{1}{4}t\int_0^t 4K^2\int_U |\beta(X_s^i, V_s, u) - \beta(\tilde{X}_s^i, \tilde{V}_s, u)|^2\mu(du)ds$$
$$\le K^4 t\int_0^t \left(|X_s^i - \tilde{X}_s^i|^2 + \rho(V_s^+, \tilde{V}_s^+)^2 + \rho(V_s^-, \tilde{V}_s^-)^2\right) ds.$$

Further, by Doob's inequality and Fatou's lemma, we can estimate the expectations of I_t^1 and I_t^2 as follows:

$$\mathbb{E}I_t^1 \le \liminf_{n\to\infty}\frac{4}{n}\sum_{i=1}^n \mathbb{E}\int_0^t |\gamma(X_s^i, V_s) - \gamma(\tilde{X}_s^i, \tilde{V}_s)|^2 1_{s\le\eta_m} ds$$
$$\le 40m^2K^2 \liminf_{n\to\infty}\frac{1}{n}\sum_{i=1}^n \int_0^t \mathbb{E}\Big(|X_s^i - \tilde{X}_s^i|^2$$
$$+ \rho(V_s^+, \tilde{V}_s^+)^2 + \rho(V_s^-, \tilde{V}_s^-)^2\Big)1_{s\le\eta_m} ds$$
$$\le 40m^2K^2\int_0^t (f_m(s) + 2m^2 f_m(s) + 2g_m(s))\, ds,$$

and

$$\mathbb{E}I_t^2 \le \liminf_{n\to\infty}\frac{4}{n}\sum_{i=1}^n \mathbb{E}\int_0^t\int_U |\beta(X_s^i, V_s, u) - \beta(\tilde{X}_s^i, \tilde{V}_s, u)|^2\mu(du)1_{s\le\eta_m} ds$$
$$\le 40m^2K^2\int_0^t (f_m(s) + 2m^2 f_m(s) + 2g_m(s))\, ds.$$

Plugging back into (4.2.7) and taking expectations, we see that there exists a constant $K_1(m,T)$,

$$g_m(t) \leq K_1(m,T) \int_0^t \left(f_m(s) + 2m^2 f_m(s) + 2g_m(s) \right) ds. \tag{4.2.8}$$

Adding (4.2.6) and (4.2.8), for $t \leq T$, we have

$$f_m(t) + g_m(t) \leq K_2(m,T) \int_0^t \left(f_m(s) + g_m(s) \right) ds, \tag{4.2.9}$$

where $K_2(m,T)$ is a constant. By Gronwall's inequality, we have

$$f_m(t) + g_m(t) = 0.$$

Then for each m and $t \in [0,T]$, we have

$$X^i_{t\wedge\eta_m} = \tilde{X}^i_{t\wedge\eta_m} \text{ and } A^i_{t\wedge\eta_m} = \tilde{A}^i_{t\wedge\eta_m}, \quad \text{a.s.}$$

Further, by (4.1.2),

$$V_{t\wedge\eta_m} = \tilde{V}_{t\wedge\eta_m} \qquad \text{a.s.}$$

Hence

$$(X_t, A_t, V_t) = (\tilde{X}_t, \tilde{A}_t, \tilde{V}_t) \qquad \text{for } t \leq \eta_m \wedge T.$$

Taking $T,\ m \to \infty$,

$$(X_t, A_t, V_t) = (\tilde{X}_t, \tilde{A}_t, \tilde{V}_t) \qquad \text{for } t \leq \eta_\infty.$$

We have by the definition of η_m,

$$\begin{aligned}
P(\eta_m \leq t) &\leq P\left(\sup_{0\leq s\leq t} \lim_{n\to\infty} \frac{1}{n} \sum_{i=1}^n |A^i_s|^2 \geq m^2 \right) \\
&\leq \frac{1}{m^2} \mathbb{E} \sup_{0\leq s\leq t} \lim_{n\to\infty} \frac{1}{n} \sum_{i=1}^n |A^i_s|^2 \\
&\leq \frac{1}{m^2} \liminf_{n\to\infty} \frac{1}{n} \sum_{i=1}^n \mathbb{E} \sup_{0\leq s\leq t} |A^i_s|^2 \\
&= \frac{1}{m^2} \mathbb{E} \sup_{0\leq s\leq t} |A^1_s|^2
\end{aligned}$$

where the last inequality follows by moving the sup inside the sum and applying Fatou's lemma, and the equality follows from the exchangeability. Hence, by Proposition 4.2.1,

$$P(\eta_\infty \leq t) = \lim_{m\to\infty} P\{\eta_m \leq t\} = 0,$$

i.e., $\eta_\infty = \infty$, a.s., and uniqueness follows. □

Finally, we establish the existence of a solution.

Theorem 4.2.3. *Under Assumptions (S1), (S2) and (4.2.3), the system has a solution.*

Proof. Define

$$B_i^n(t) = B_i\left(\frac{[nt]}{n}\right), \qquad D_n(t) = \frac{[nt]}{n},$$

and

$$W^n(B \times [0,t]) = W\left(B \times \left[0, \frac{[nt]}{n}\right]\right), \qquad \forall B \in \mathcal{U}.$$

Consider the discrete time, Euler-type approximation $(X^{i,n}, A^{i,n})$ obtained by replacing B_i by B_i^n, dt by $dD_n(t)$, and W by W^n in (4.1.3) and defining

$$A_t^{i,n} = A_0^i \exp\Big\{\int_0^t \gamma(X_{s-}^{i,n}, V_{s-}^n)dB_i^n(s) + \int_0^t D(X_{s-}^{i,n}, V_{s-}^n)dD_n(s) + \int_0^t \int_U \beta(X_{s-}^{i,n}, V_{s-}^n, u)W^n(duds)\Big\},$$

where

$$D(x,\nu) = d(x,\nu) - \frac{1}{2}|\gamma(x,\nu)|^2 - \frac{1}{2}\int_U \beta(x,\nu,u)^2\mu(du).$$

Note that the exchangeability of $\{(X_0^i, A_0^i)\}_{i=1}^\infty$ gives the existence of $V_t^n = V_0^n$ for $0 \le t < \frac{1}{n}$ and the exchangeability of $\left\{\left(X_{1/n}^{i,n}, A_{1/n}^{i,n}\right)\right\}_{i=1}^\infty$. The exchangeability of $\{(X_t^{i,n}, A_t^{i,n})\}_{i=1}^\infty$ and the existence of V_t^n then follows recursively.

Let

$$K_1 = \sup_{x,v} |c(x,v)|$$

and

$$K_2 = \sup_{x,v}\left(|\sigma(x,v)\sigma(x,v)| + \int_U |\alpha(x,v,u)|^2\mu(du)\right).$$

Then

$$\mathbb{E}[|X_{t+h}^{i,n} - X_t^{i,n}|^2|\mathcal{F}_t] \le 2\left(K_1\left(\frac{[n(t+h)]-[nt]}{n}\right)^2 + K_2\frac{[n(t+h)]-[nt]}{n}\right),$$

with a similar estimate holding for $\log|A^{i,n}|$. By Theorem 3.8.6 and Remark 3.8.7 of Ethier and Kurtz (1986), for each i, $\{(X^{i,n}, A^{i,n})\}$ is relatively compact for convergence in distribution in $D(\mathbb{R}_+, \mathbb{R} \times \mathbb{R})$. But relative compactness of $\{(X^{i,n}, A^{i,n})\}$ in $D(\mathbb{R}_+, \mathbb{R}\times\mathbb{R})$ implies relative compactness of $\{(X^n, A^n)\}$ in $D(\mathbb{R}_+, \mathbb{R} \times \mathbb{R})^\infty$. (See, for example, Ethier and Kurtz (1986), Proposition 3.2.4.)

Taking a subsequence if necessary, we assume that $(X^n, A^n) \Rightarrow (X, A)$. By the continuity of B_i and W and the boundedness of the coefficients in (4.1.3) and (4.1.4), (X^i, A^i) will be continuous for each i, and it follows that the convergence is, in fact, in $D(\mathbb{R}_+, (\mathbb{R} \times \mathbb{R})^\infty)$. Define

$$Z_t^n = \lim_{m\to\infty} \frac{1}{m} \sum_{i=1}^m \delta_{(X_t^{i,n}, A_t^{i,n})}, \quad Z_t = \lim_{m\to\infty} \frac{1}{m} \sum_{i=1}^m \delta_{(X_t^i, A_t^i)}.$$

Then by Theorem A.2.1, $Z^n \Rightarrow Z$, or more precisely, $(X^n, A^n, Z^n) \Rightarrow (X, A, Z)$.

For simplicity, assume that $A_0^i \geq 0$ for all i. If, for $\alpha > 0$, we define $V^{n,\alpha}$ by

$$\langle V^{n,\alpha}, \varphi \rangle = \int_{\mathbb{R}^2} (a \wedge \alpha)\varphi(x) Z^n(t, dxda),$$

and observe that

$$\begin{aligned} \|V_t^n - V_t^{n,\alpha}\| &= \lim_{m\to\infty} \frac{1}{m} \sum_{i=1}^m (A_t^{i,n} - \alpha \wedge A_t^{i,n}) \\ &\leq \lim_{m\to\infty} \frac{1}{m} \sum_{i=1}^m \left(\sup_{s\leq T} A_s^{i,n} - \alpha \wedge \sup_{s\leq T} A_s^{i,n} \right). \end{aligned}$$

By the same argument as in the proof of Proposition 4.2.1, $\{\sup_{t\leq T} A_t^{1,n}\}$ is bounded in L^2 and hence uniformly integrable, so

$$\mathbb{E} \sup_{t\leq T} \|V_t^n - V_t^{n,\alpha}\| \leq \mathbb{E} \left(\sup_{t\leq T} A_t^1 - \alpha \wedge \sup_{t\leq T} A_t^1 \right). \tag{4.2.10}$$

Since the right hand side of (4.2.10) goes to zero as $\alpha \to \infty$, it follows that (X^n, A^n, V^n) is relatively compact, and as in Theorem A.2.1, any limit point will be a distributional solution of (4.1.3)-(4.1.4). But as in Yamada and Watanabe (1971) (cf. Theorem A.3.2), distributional existence and pathwise uniqueness imply strong existence. □

In the classical setting, the limiting empirical process is deterministic and characterized by a McKean-Vlasov equation. Here the limiting equation (4.1.6) is still stochastic due to the effect of the common random measure W.

The classical McKean-Vlasov limit (without W or weights) is sometimes described by the equation

$$X_t = X_t + \int_0^t \sigma(X_s, Z_s)dB(s) + \int_0^t c(X_s, Z_s)ds,$$

where Z_t is required to be the distribution of X_t. The analogous formulation in our setting is to consider the system

$$X_t = X_t + \int_0^t \sigma(X_s, V_s)dB(s) + \int_0^t c(X_s, V_s)ds \qquad (4.2.11)$$
$$+ \int_0^t \int_U \alpha(X_s, V_s, u)W(dsdu)$$

and

$$A_t = A_t + \int_0^t A_t\gamma(X_s, V_s)dB(s) + \int_0^t A_t d(X_s, V_s)ds \qquad (4.2.12)$$
$$+ \int_0^t \int_U A_t\beta(X_s, V_s, u)W(dsdu),$$

where, as we will see below, V_t is the random measure determined by

$$\langle V_t, \phi \rangle = \mathbb{E}(A_t\phi(X_t)|\mathcal{F}_t^W), \qquad (4.2.13)$$

$\{\mathcal{F}_t^W\}$ being the filtration generated by W. We require (X, A) to be compatible with (B, W) in the sense that for each time $t \geq 0$, the increments of B and W after time t are independent of $\mathcal{F}_t^{X,A,B,W}$. Note that this independence implies (4.2.13). As a characterization of V, this system is essentially equivalent to the particle system.

Theorem 4.2.4. *Let (X, A, V, B, W) satisfy (4.2.11)-(4.2.13). Then there exists a solution*

$$(\{X^i\}, \{A^i\}, \{B_i\}, \tilde{V}, \tilde{W})$$

of (4.1.2)-(4.1.4) such that $(X^1, A^1, \tilde{V}, B_1, \tilde{W})$ has the same distribution as (X, A, V, B, W). Conversely, if there exists a pathwise unique solution $(\{X^i\}, \{A^i\}, \{B_i\}, V, W)$ of (4.1.2)-(4.1.4), then (X^1, A^1, V, B_1, W) is a solution of (4.2.11)-(4.2.13).

Proof. Since we are not assuming uniqueness, (X, A) may not be uniquely determined by (X_0, A_0, B, W); however, if we let (X, A, V, B, W) be a particular solution of (4.2.11)-(4.2.13), then (X, A) will have a regular conditional distribution given (X_0, A_0, B, W). In particular, there will exist a transition function $q(x_0, a_0, b, w, \Gamma)$ such that

$$P\{(X, A) \in \Gamma | X_t, A_t, B, W\} = q(X_t, A_t, B, W, \Gamma),$$

for any $\Gamma \in \mathcal{B}(D(\mathbb{R}_+, \mathbb{R} \times \mathbb{R}))$. Since every probability measure on a complete, separable metric space can be induced by a mapping from the probability space given by the Lebesgue measure on $[0, 1]$, it follows that there will be a mapping F such that if ξ is uniformly distributed on the interval $[0, 1]$ and $(\tilde{X}_0, \tilde{A}_0, \tilde{B}, \tilde{W})$ is independent of ξ and has the same distribution as (X_0, A_0, B, W), then

$$(\tilde{X}, \tilde{A}) = F(\tilde{X}_0, \tilde{A}_0, \tilde{B}, \tilde{W}, U)$$

and $(\tilde{X}_0, \tilde{A}_0, \tilde{B}, \tilde{W})$ have the same joint distribution as (X, A) and (X_0, A_0, B, W). Defining $\tilde{V}$ by

$$\left\langle \tilde{V}_t, \phi \right\rangle = \mathbb{E}(\tilde{A}_t \phi(\tilde{X}_t) | \mathcal{F}_t^{\tilde{W}}), \tag{4.2.14}$$

$(\tilde{X}, \tilde{A}, \tilde{V}, \tilde{B}, \tilde{W})$ will have the same distribution as (X, A, V, B, W). Let W be a white noise random measure, $\{B_i\}$ be independent standard Brownian motions, $\{(X_0^i, A_0^i)\}_{i=1}^\infty$ be i.i.d. with the same distribution as (X_0, A_0), and $\{\xi_i\}_{i=1}^\infty$ be independent uniform-$[0, 1]$ random variables. Define

$$(X^i, A^i) = F(X_0^i, A_0^i, B_i, W, \xi_i).$$

Note that V_t determined by

$$\langle V_t, \phi \rangle = \mathbb{E}(A_t^i \phi(X_t^i) | \mathcal{F}_t^W),$$

does not depend on i and that $(\{X^i\}, \{A^i\}, V, \{B_i\}, W)$ satisfies (4.1.2), (4.1.3) and (4.1.4). It remains only to show that V satisfies (4.1.2).

Note that $\{(X^i, A^i)\}_{i=1}^\infty$ is exchangeable so

$$\left\langle \tilde{V}_t, \phi \right\rangle = \lim_{n \to \infty} \frac{1}{n} \sum_{i=1}^n A_t^i \phi(X_t^i) = E(A_t^1 \phi(X_t^1) | \mathcal{I})$$

exists. The second equality holds by the ergodic theorem, and $\mathcal{I}$ is the invariant σ-algebra for the stationary sequence $\{(X_0^i, A_0^i, B_i, \xi_i, W)\}_{i=1}^\infty$. But the independence of $\{(X_0^i, A_0^i, B_i, \xi_i)\}$ implies $\mathcal{I}$ is contained in the completion of the σ-algebra generated by W. Consequently,

$$\left\langle \tilde{V}_t, \phi \right\rangle = E(A_t^1 \phi(X_t^1) | W) = E(A_t^1 \phi(X_t^1) | \mathcal{F}_t^W),$$

where the second equality follows by (4.2.13), and hence $V_t = \tilde{V}_t$.

To obtain the converse, note that pathwise uniqueness implies that the invariant σ-algebra for $\{(X^i, A^i, B_i, W)\}$ is contained in the completion of $\sigma(W)$. It can also be concluded from pathwise uniqueness that the solution $\{(X^i, A^i)\}$ is compatible with the $\{B_i\}$ and W, so we have

$$\langle V_t, \phi \rangle = E(A_t^1 \phi(X_t^1) | W) = E(A_t^1 \phi(X_t^1) | \mathcal{F}_t^W).$$

□

4.3 A nonlinear SPDE

In this section, we establish the existence and uniqueness for the solution to the nonlinear SPDE (4.1.6). We begin with a summary of the techniques used in this section. First, by applying Itô's formula, it is shown that V is a solution to (4.1.6). To prove uniqueness for the solution to (4.1.6), we assume the existence of another solution V^1 and freeze the nonlinear arguments in (4.1.6) by V^1 (cf. (4.3.4) and (3.1.1)) to obtain a linear SPDE. Similar to the argument in Xiong (1995), the uniqueness for the solution to (4.1.6) is implied by that of the linear SPDE (4.3.4) and that of the system (4.1.2)-(4.1.4) proved in the previous section.

We actually only prove uniqueness among solutions V such that for each $t \geq 0$, V_t is absolutely continuous with respect to the Lebesgue measure and has a density in $L^2(\mathbb{R})$ (we also prove existence of such a solution for all V_0 with this property).

Theorem 4.3.1. *Let V be the weighted empirical measure for the particle system given by Theorems 4.2.2 and 4.2.3. Then V is a solution to SPDE (4.1.6).*

Proof. It is easy to see that

$$\begin{aligned}
&\mathbb{E}\sup_{t\leq T}\left|\frac{1}{n}\sum_{i=1}^{n}\int_0^t A_s^i\nabla\phi(X_s^i)\sigma(X_s^i,V_s)dB_i(s)\right|^2\\
\leq\ & 4\frac{1}{n^2}\sum_{i=1}^{n}\mathbb{E}\int_0 |A_s^i|^2\left|\nabla\phi(X_s^i)\sigma(X_s^i,V_s)\right|^2 ds\\
\leq\ & \frac{4}{n}\|\nabla\phi\|_\infty^2 K^2 T\mathbb{E}\sup_{s\leq T}|A_s^1|^2 \to 0.
\end{aligned}$$

By (4.1.5), it is then easy to prove that V is a solution of (4.1.6). □

Finally, we consider the uniqueness of the solution of the nonlinear SPDE (4.1.6).

Theorem 4.3.2. *Suppose that $V_0 \in H_0$, then there exists a unique H_0-valued solution of (4.1.6).*

Proof. Let V be the solution of (4.1.6) given by Theorem 4.3.1. Then by Corollary 3.1.5, V^+ and V^- (and hence V) have values in H_0.

Let V_t^1 be another H_0-valued solution of (4.1.6). Consider the system of SDEs: $i = 1, 2, \cdots$,

$$X_t^i = X_0^i + \int_0^t \sigma(X_s^i, V_s^1) dB_i(s) + \int_0^t c(X_s^i, V_s^1) ds \qquad (4.3.1)$$
$$+ \int_0^t \int_U \alpha(X_s^i, V_s^1, u) W(dsdu)$$

and

$$A_t^i = A_0^i + \int_0^t A_s^i \gamma(X_s^i, V_s^1) dB_i(s) + \int_0^t A_s^i d(X_s^i, V_s^1) ds \qquad (4.3.2)$$
$$+ \int_0^t \int_U A_s^i \beta(X_s^i, V_s^1, u) W(dsdu).$$

Let $V_t^{2,\pm}$ be given by

$$V_t^{2,\pm} = \lim_{n\to\infty} \frac{1}{n} \sum_{i=1}^{\infty} A_t^{i,\pm} \delta_{X_t^i}. \qquad (4.3.3)$$

As in Theorem 4.3.1, $V^{2,+}$ and $V^{2,-}$ are solutions of

$$\langle U_t, \phi \rangle = \langle U_0, \phi \rangle + \int_0^t \left\langle U_s, d(\cdot, V_s^1)\phi + L_{V_s^1}\phi \right\rangle ds \qquad (4.3.4)$$
$$+ \int_0^t \int_U \left\langle U_s, \beta(\cdot, V_s^1, u)\phi + \alpha^T(\cdot, V_s^1, u)\nabla\phi \right\rangle W(dsdu).$$

By Corollary 3.1.5, $V^{2,+}$ and $V^{2,-}$ (and hence $V^2 = V^{2,+} - V^{2,-}$) are H_0-valued. In particular, V^2 is an H_0-valued solution to the linear SPDE (4.3.4). Since V^1 is also an H_0-valued solution of (4.3.4), it follows from Theorem 3.1.7 that $V^2 = V^1$. Hence, V^1 corresponds to a solution of the system (4.1.2)-(4.1.4). By the uniqueness of the solution to this system we see that $V_t = V_t^1$. □

4.4 Historical remarks

Limits of empirical measure processes for systems of interacting diffusions have been studied by various authors (see, for example, Chiang, Kallianpur and Sundar (1991), Graham (1992), Hitsuda and Mitoma (1986), Kallianpur and Xiong (1994), Méléard (1996), and Morien (1996)) since the pioneering work by McKean (1967). Typically, the driving processes in the models are assumed to be independent, which makes the limit become a deterministic measure-valued function.

Florchinger and Le Gland (1992) considered particle approximations for stochastic partial differential equations in a setting that, in the notation in this chapter, corresponds to taking $\gamma = \sigma = 0$ and letting the other coefficients be independent of V. Florchinger and Le Gland (1992) were motivated by approximations to the Zakai equation of nonlinear filtering. Del Moral (1995) specifically studied this example. Kotelenez (1995) introduced a model of n-particles with the same driving process for each particle and studied the empirical process as the solution of an SPDE. His model corresponds to taking $\gamma = \sigma = d = \beta = 0$, but the other coefficients are allowed to depend on V. In particular, the weights A^i are constants. Dawson and Vaillancourt (1995) consider a model given as a solution of a martingale problem that corresponds to taking $A_t^i \equiv 1$ in the current model. Bernard, Talay, and Tubaro (1994) investigated a system with time-varying weights and a deterministic limit.

The material of this chapter is taken from Kurtz and Xiong (1999). The idea of the decomposition of the uniqueness problem for a nonlinear SPDE to that of a linear SPDE and an infinite particle system is motivated by that of Xiong (1995) where a nonlinear PDE was treated.

Chapter 5

Stochastic Log-Laplace Equation

5.1 Introduction

As we indicated in Chapter 1, log-Laplace equation has been used by many authors in deriving various properties for the superprocesses. For example, it can be used to prove the branching property of the superprocess. A counterpart of this equation for superprocesseses in random environment is the stochastic log-Laplace equation (SLLE) which we will study in this chapter.

First, we write the CMP introduced in Chapter 2 formally as: For any $f \in C_0^2(\mathbb{R})$,

$$N_t(f) \equiv \langle X_t, f \rangle - \langle \nu, f \rangle - \int_0^t \langle X_s, L_s^W f \rangle \, ds$$

is a continuous P^W-martingale with quadratic variation process

$$\langle N(f) \rangle_t = \gamma \int_0^t \langle X_s, f^2 \rangle \, ds,$$

where

$$L_s^W f(x) = \int_U h(y,x) \dot{W}_{sy} dy f'(x) + \frac{1}{2} a(x) f''(x).$$

Here, formally, $\dot{W}_{sy}$ denotes the "derivative" of the random measure W, i.e., it is the space-time white noise.

Similar to Theorem 1.2.2, we then "must have" that for any $f \in C_b^+(\mathbb{R})$,

$$\mathbb{E}^W \exp\left(-\langle X_t, f \rangle\right) = \exp\left(-\langle \nu, v_{0,t} \rangle\right), \tag{5.1.1}$$

where, for fixed t, $\{v_{s,t} : s \in [0,t]\}$ is the unique solution to the following backward SPDE

$$\begin{cases} \frac{\partial}{\partial s} v_{s,t}(x) = -L v_{s,t}(x) - \int_U h(y,x) \dot{W}_{sy} dy \nabla v_{s,t}(x) + \frac{\gamma}{2} v_{s,t}(x)^2 \\ v_{t,t}(x) \quad = f(x). \end{cases} \tag{5.1.2}$$

Recall that L is the differential operator given by

$$Lf(x) = \frac{1}{2}a(x)f''(x), \qquad \forall\, x \in \mathbb{R}.$$

Note that (5.1.2) is understood in a formal writing only. To make it rigorous, we consider its integral form:

$$v_{s,t}(x) = f(x) + \int_s^t \left(Lv_{r,t}(x) - \frac{\gamma}{2} v_{r,t}(x)^2 \right) dr + \int_s^t \int_U h(y,x)\nabla v_{r,t}(x) W(\hat{dr}dy), \tag{5.1.3}$$

where notation $W(\hat{dr}dy)$ means that the stochastic integral is the backward Itô integral introduced in Chapter 3. Namely, when we define the stochastic integral by Riemann sum approximation, we use the right end points instead of the left ones. Note that this is the essential difference between backward SPDE and backward PDE. For the latter, the backward integral is the same as the forward one.

For the simplicity of notation, we consider the forward version of SPDE (5.1.3):

$$v_t(x) = f(x) + \int_0^t \left(Lv_r(x) - \frac{\gamma}{2} v_r(x)^2 \right) dr + \int_0^t \int_U h(y,x)\nabla v_r(x) W(drdy). \tag{5.1.4}$$

This is the first class of nonlinear SPDEs for the SPREs we shall study in this monograph. We will establish the existence and uniqueness of the solution to (5.1.4). We will also present its particle system representation of the form introduced in Chapter 4.

Recall that for integer m,

$$H_m = \{\phi \in H_0 : \phi^{(j)} \in H_0, \ j \le m\},$$

where $\phi^{(j)}$ stands for the jth derivative of ϕ. We define Sobolev norm on H_m by

$$\|\phi\|_m^2 = \sum_{j=0}^m \int_{\mathbb{R}} |\phi^{(j)}(x)|^2 dx.$$

Recall that $\langle \cdot,\cdot \rangle_0$ denotes the inner product in H_0.

Definition 5.1.1. An H_0^+-valued (measurable) process v_t is a solution to (5.1.4) if for any $\phi \in C_0^\infty(\mathbb{R})$,

$$\langle v_t, \phi \rangle_0 = \langle f, \phi \rangle_0 + \int_0^t \langle v_r, L^*\phi - v_r\phi \rangle_0 \, dr - \int_0^t \int_U \langle v_r, \nabla(h(y,\cdot)\phi) \rangle_0 \, W(drdy), \qquad t \ge 0.$$

Throughout this chapter, we assume the following assumption:

Boundedness condition (BC): Suppose that $f \geq 0$, $a \geq 0$ and h taking values in $L^2(U, \mathcal{U}, \mu)$ are bounded functions with bounded first and second order derivatives.

Note that the derivatives of h always mean those with respect to the second variable. Namely, those of $h(y,x)$ with respect to x.

The following theorem is the main result of this chapter whose proof will be presented in the next few sections.

Theorem 5.1.2. *Suppose that condition (BC) holds. Then,*
i) The SLLE (5.1.4) has a unique solution $v_t(x)$.
ii) v_t is the unique solution of the following infinite particle system: $i = 1, 2, \cdots$,

$$\begin{aligned} d\xi_t^i = dB_i(t) - \int_U h(y, \xi_t^i) W(dtdy) \\ + \left(2a'(\xi_t^i) - \int_U h(y, \xi_t^i) \nabla h(y, \xi_t^i) \mu(dy) \right) dt, \end{aligned} \tag{5.1.5}$$

$$dm_t^i = m_t^i \left(\left(a''(\xi_t^i) - v_t(\xi_t^i) \right) dt - \int_U \nabla h(y, \xi_t^i) W(dtdy) \right), \tag{5.1.6}$$

$$\nu_t = \lim_{n\to\infty} \frac{1}{n} \sum_{i=1}^n m_t^i \delta_{\xi_t^i}, \qquad a.s. \tag{5.1.7}$$

where for any $t \geq 0$, ν_t is absolutely continuous with respect to Lebesgue measure with v_t as the Radon-Nikodym derivative.

5.2 Approximation and two estimates

To establish the existence of a nonnegative solution to the nonlinear SPDE (5.1.4), we smooth and truncate its nonlinear term and consider

$$\begin{aligned} v_t^\epsilon(x) = f(x) + \int_0^t \left(Lv_r^\epsilon(x) - \frac{\gamma}{2} v_r^\epsilon(x) T_\epsilon v_r^{\epsilon,\epsilon}(x) \right) dr \\ + \int_0^t \int_U h(y, x) \nabla v_r^\epsilon(x) W(drdy) \end{aligned} \tag{5.2.1}$$

where T_ϵ is the Brownian semigroup defined in Chapter 1,

$$v_r^{\epsilon,\epsilon}(x) = \lambda_\epsilon(v_r^\epsilon) v_r^\epsilon(x)$$

and $\lambda_\epsilon : L^1_+(\mathbb{R}) \to \mathbb{R}_+$ is given by

$$\lambda_\epsilon(f) = \frac{\int_{\mathbb{R}} f(x)dx \wedge \epsilon^{-1}}{\int_{\mathbb{R}} f(x)dx}$$

with the convention that $\frac{0}{0} = 0$. Here, $L^1_+(\mathbb{R})$ is the collection of all non-negative integrable functions on $\mathbb{R}$.

Lemma 5.2.1. *For every $\epsilon > 0$, the approximating SPDE (5.2.1) has a unique solution.*

Proof. Consider the following infinite particle system: $i = 1, 2, \cdots$,

$$\begin{cases} d\xi^i_t &= dB_i(t) - \int_U h(y, \xi^i_t) W(dtdy) \\ &\quad + \left(2a'(\xi^i_t) - \int_U h(y, \xi^i_t) \nabla h(y, \xi^i_t) \mu(dy)\right) dt, \\ dm^{\epsilon,i}_t &= m^{\epsilon,i}_t \left(\left(a''(\xi^i_t) - \frac{\gamma}{2} T_\epsilon \nu^{\epsilon,\epsilon}_t(\xi^i_t)\right) dt - \int_U \nabla h(y, \xi^i_t) W(dtdy) \right) \\ \nu^\epsilon_t &= \lim_{n\to\infty} \frac{1}{n} \sum_{i=1}^n m^{\epsilon,i}_t \delta_{\xi^i_t} \qquad \text{a.s.,} \end{cases} \tag{5.2.2}$$

where $\nu^{\epsilon,\epsilon} = \lambda_\epsilon(\nu^\epsilon)\nu^\epsilon$ and, we abuse the notation a bit by using the notation λ_ϵ for the mapping from $\mathcal{M}_F(\mathbb{R})$ to $\mathbb{R}_+$ defined as

$$\lambda_\epsilon(\nu) = \frac{\nu(\mathbb{R}) \wedge \epsilon^{-1}}{\nu(\mathbb{R})}, \qquad \forall\, \nu \in \mathcal{M}_F(\mathbb{R}).$$

Recall that we also use T_t as a mapping from $\mathcal{M}_F(\mathbb{R})$ to $C_b(\mathbb{R})$ given by

$$T_t\nu(x) = \int_{\mathbb{R}} p_t(x-y)\nu(dy).$$

Now we show that the conditions (S1) and (S2) of Chapter 4 are satisfied by the coefficients of the system (5.2.2). We only check those for

$$d_\epsilon(x, \nu) \equiv -\lambda_\epsilon(\nu) T_\epsilon \nu(x).$$

The verification for other coefficients is trivial.

Note that $p_\epsilon(x) \le (2\pi\epsilon)^{-1/2}$. Then,

$$|d_\epsilon(x,\nu)| = \lambda_\epsilon(\nu) \left| \int_{\mathbb{R}} p_\epsilon(x-y)\nu(dy) \right| \le (2\pi\epsilon)^{-1/2}\epsilon^{-1}.$$

Also note that

$$\begin{aligned} |\nabla p_\epsilon(x)| &\le (2\pi)^{-1/2}\epsilon^{-1} \sup_x e^{-\frac{|x|^2}{2\epsilon}} \frac{|x|}{\sqrt{\epsilon}} \\ &= (2\pi)^{-1/2}\epsilon^{-1} e^{-1/2} \equiv K_\epsilon. \end{aligned}$$

Therefore,

$$\left| \int_{\mathbb{R}} (p_\epsilon(x_1 - y) - p_\epsilon(x_2 - y))\nu_1(dy) \right| \le K_\epsilon |x_1 - x_2| \nu_1(\mathbb{R}).$$

We recall the Wasserstein distance on $\mathcal{M}_F(\mathbb{R})$ defined as

$$\rho(\nu_1, \nu_2) = \sup_{g \in \mathbb{B}_1} |\langle \nu_1 - \nu_2, g\rangle|,$$

where

$$\mathbb{B}_1 = \{g \in C(\mathbb{R}) : \ |g(x)| \leq 1, \quad |g(x) - g(y)| \leq |x - y|, \ \forall \ x, \ y \in \mathbb{R}\}.$$

For $g \in \mathbb{B}_1$, we have

$$\begin{aligned} |\lambda_\epsilon(\nu_1) - \lambda_\epsilon(\nu_2)| &= \left|\frac{\nu_1(\mathbb{R}) \wedge \epsilon^{-1}}{\nu_1(\mathbb{R})} - \frac{\nu_2(\mathbb{R}) \wedge \epsilon^{-1}}{\nu_2(\mathbb{R})}\right| \\ &\leq \nu_1(\mathbb{R})^{-1} \left|\nu_1(\mathbb{R}) \wedge \epsilon^{-1} - \nu_2(\mathbb{R}) \wedge \epsilon^{-1}\right| \\ &\quad + \nu_2(\mathbb{R}) \wedge \epsilon^{-1} \left|\nu_1(\mathbb{R})^{-1} - \nu_2(\mathbb{R})^{-1}\right| \\ &\leq \nu_1(\mathbb{R})^{-1} \left|\nu_1(\mathbb{R}) - \nu_2(\mathbb{R})\right| + \nu_1(\mathbb{R})^{-1} \left|\nu_1(\mathbb{R}) - \nu_2(\mathbb{R})\right| \\ &\leq 2\nu_1(\mathbb{R})^{-1} \rho(\nu_1, \nu_2). \end{aligned}$$

Then,

$$\begin{aligned} &|d_\epsilon(x_1, \nu_1) - d_\epsilon(x_2, \nu_2)| \\ &\leq \left|\int_{\mathbb{R}} (p_\epsilon(x_1 - y) - p_\epsilon(x_2 - y))\nu_1(dy)\right| \lambda_\epsilon(\nu_1) \\ &\quad + \int_{\mathbb{R}} p_\epsilon(x_2 - y)\nu_1(dy) |\lambda_\epsilon(\nu_1) - \lambda_\epsilon(\nu_2)| \\ &\quad + \lambda_\epsilon(\nu_2) \left|\int_{\mathbb{R}} p_\epsilon(x_2 - y)\nu_1(dy) - \int_{\mathbb{R}} p_\epsilon(x_2 - y)\nu_2(dy)\right| \\ &\leq K_\epsilon \epsilon^{-1} |x_1 - x_2| + \frac{2}{\sqrt{2\pi\epsilon}} \rho(\nu_1, \nu_2) + + \left((2\pi\epsilon)^{-1/2} \wedge K_\epsilon\right) \rho(\nu_1^\epsilon, \nu_2^\epsilon) \\ &\leq K_1 \left(|x_1 - x_2| + \rho(\nu_1, \nu_2)\right), \end{aligned}$$

where K_1 is a constant which may depend on ϵ.

By Theorems 4.3.1 and 4.3.2, ν_t^ϵ is the unique solution to

$$\begin{aligned} \langle \nu_t^\epsilon, \phi\rangle = \langle f, \phi\rangle + \int_0^t \left\langle \nu_r^\epsilon, L^*\phi - \frac{\gamma}{2}(T_\epsilon \nu_r^{\epsilon,\epsilon})\phi\right\rangle dr \\ - \int_0^t \int_U \langle \nu_r^\epsilon, \nabla(h(y, \cdot)\phi)\rangle W(drdy). \end{aligned}$$

Further, ν_t^ϵ is absolutely continuous with respect to Lebesgue measure with Radon-Nikodym derivative v_t^ϵ which belongs to H_0 and solves SPDE (5.2.1).

□

Next, we establish a comparison result for SPDEs of the form (5.2.1). As a consequence, we obtain the boundedness of v_t^ϵ. We will also give a Sobolev type estimate.

Lemma 5.2.2. *For all $r \ge 0$ and $x \in \mathbb{R}$, we have*

$$v_r^\epsilon(x) \le \|f\|_\infty, \qquad a.s.$$

where $\|f\|_\infty$ is the supremum norm of f.

Proof. Let $\tilde{m}_t^i$ be given by

$$d\tilde{m}_t^i = \tilde{m}_t^i \left(a''(\xi_t^i)dt - \int_U \nabla h(y, \xi_t^i) W(dtdy) \right)$$

and let

$$\tilde{\nu}_t = \lim_{n\to\infty} \frac{1}{n} \sum_{i=1}^n \tilde{m}_t^i \delta_{\xi_t^i}, \qquad \text{a.s.}$$

Then $m_t^{\epsilon,i} \le \tilde{m}_t^i$ and hence, for $\phi \ge 0$,

$$\langle \nu_t^\epsilon, \phi \rangle \le \langle \tilde{\nu}_t, \phi \rangle . \tag{5.2.3}$$

Similar to lemma 5.2.1, it is easy to show that

$$\langle \tilde{\nu}_t, \phi \rangle = \langle f, \phi \rangle + \int_0^t \langle \tilde{\nu}_r, L^*\phi \rangle \, dr - \int_0^t \int_U \langle \tilde{\nu}_r, \nabla(h(y,\cdot)\phi) \rangle \, W(drdy). \tag{5.2.4}$$

Let ϕ_t be given as a solution of the linear SPDE

$$\langle \phi_t, f \rangle = \langle \phi, f \rangle + \int_0^t \langle \phi_r, Lf \rangle \, dr + \int_0^t \int_U \langle \phi_r, h(y,\cdot) f' \rangle \, \tilde{W}(drdy) \tag{5.2.5}$$

where $\tilde{W}$ is an independent copy of W. The existence of a solution to (5.2.5) follows from Theorem 2.3.2 with $\gamma = 0$. By Itô's formula, we see that

$$e^{-\alpha\langle \tilde{\nu}_t, \phi \rangle} - \int_0^t e^{-\alpha\langle \tilde{\nu}_s, \phi \rangle} \left(\alpha \langle \tilde{\nu}_s, L^*\phi \rangle - \frac{\alpha^2}{2} \int_U \langle \tilde{\nu}_r, \nabla(h(y,\cdot)\phi) \rangle^2 \mu(dy) \right) ds$$

and

$$e^{-\alpha\langle \phi_t, f \rangle} - \int_0^t e^{-\alpha\langle f, \phi_s \rangle} \left(\alpha \langle Lf, \phi_s \rangle - \frac{\alpha^2}{2} \int_U \langle h(y,\cdot) f', \phi_r \rangle^2 \mu(dy) \right) ds$$

are martingales. By the duality argument (Theorem 1.3.1), we have

$$\mathbb{E} e^{-\alpha\langle \tilde{\nu}_t, \phi \rangle} = \mathbb{E} e^{-\alpha\langle \phi_t, f \rangle}.$$

This implies that $\langle\tilde{\nu}_t, \phi\rangle$ and $\langle\phi_t, f\rangle$ have the same distribution. Taking $f \equiv 1$ in (5.2.5), it is clear that

$$\langle\tilde{\nu}_t, \phi\rangle = \langle 1, \phi\rangle,$$

and hence,

$$\int_{\mathbb{R}} \phi_t(x)dx = \int_{\mathbb{R}} \phi(x)dx, \qquad \text{a.s.}$$

Then,

$$\langle\phi_t, f\rangle \leq \|f\|_\infty \int_{\mathbb{R}} \phi(x)dx, \qquad \text{a.s.}$$

and hence,

$$\langle\tilde{\nu}_t, \phi\rangle \leq \|f\|_\infty \int_{\mathbb{R}} \phi(x)dx, \qquad \text{a.s.}$$

This implies the conclusion of the lemma. □

From the proof of the above lemma, we see that

Corollary 5.2.3.

$$\sup_{0\leq t\leq T} |\lambda_\epsilon(\nu_t^\epsilon) - 1| \to 0 \qquad a.s.$$

as $\epsilon \to 0$.

Proof. Taking $\phi = 1$ in (5.2.4), we get

$$\langle\tilde{\nu}_t, 1\rangle = \langle f, 1\rangle + \int_0^t \left\langle \tilde{\nu}_r, \frac{1}{2}a''\right\rangle dr - \int_0^t \int_U \langle\tilde{\nu}_r, \nabla h(y,\cdot)\rangle\, W(drdy).$$

Using Burkholder-Davis-Gundy inequality, by condition (BC), we obtain

$$\begin{aligned}\mathbb{E}\sup_{s\leq t} \langle\tilde{\nu}_s, 1\rangle^2 &\leq 3\langle f, 1\rangle^2 + K_1 \int_0^t \langle\tilde{\nu}_s, 1\rangle\, ds \\ &\quad +12\mathbb{E}\int_0^t \int_U \langle\tilde{\nu}_s, \nabla h(y,\cdot)\rangle^2 \mu(dy)ds \\ &\leq 3\langle f, 1\rangle^2 + K_2 \int_0^t \mathbb{E}\langle\tilde{\nu}_s, 1\rangle\, ds. \end{aligned} \tag{5.2.6}$$

Applying Gronwall's argument to (5.2.6), we see that

$$\mathbb{E}\sup_{s\leq T} \langle\tilde{\nu}_s, 1\rangle^2 < \infty,$$

and hence,

$$\sup_{s\leq T} \langle\tilde{\nu}_s, 1\rangle < \infty, \qquad \text{a.s.}$$

The conclusion then follows from (5.2.3) and the definition of λ_ϵ. □

Now we give an estimate for the Sobolev norm of v_t^ϵ. Recall that $\|\cdot\|_1$ is the Sobolev norm in the Hilbert space H_1.

Lemma 5.2.4. *There exists a constant K_1 such that*

$$\mathbb{E} \sup_{0\le t\le T} \|v_t^\epsilon\|_1^4 \le K_1. \tag{5.2.7}$$

Proof. We freeze the nonlinear term in (5.2.1) and consider $v_t^\epsilon(x)$ as the unique solution to the following linear equation

$$\begin{aligned} u_t^\epsilon(x) &= f(x) + \int_0^t \left(Lu_r^\epsilon(x) - \frac{\gamma}{2} u_r^\epsilon(x) T_\epsilon v_r^{\epsilon,\epsilon}(x) \right) dr \\ &\quad + \int_0^t \int_U h(y,x) \nabla u_r^\epsilon(x) W(drdy). \end{aligned} \tag{5.2.8}$$

By Rozovskii (1990), the solution has derivatives, for which estimates depend on the bounds of a, $T_\epsilon v_r^{\epsilon,\epsilon}$, h and their derivatives. Since the bound of the derivative of $T_\epsilon v_r^{\epsilon,\epsilon}$ may depend on ϵ, we *cannot* apply Rozovskii's estimate directly. Instead, we derive our estimate here. Note that

$$\begin{aligned} \langle u_t^\epsilon, \phi\rangle_0 &= \langle \phi, f\rangle_0 + \int_0^t \left\langle Lu_r^\epsilon - \frac{\gamma}{2} u_r^\epsilon T_\epsilon v_r^{\epsilon,\epsilon}, \phi \right\rangle_0 dr \\ &\quad + \int_0^t \int_U \langle h(y,\cdot) \nabla u_r^\epsilon, \phi\rangle_0 W(drdy). \end{aligned}$$

By Itô's formula, we have

$$\begin{aligned} \langle u_t^\epsilon, \phi\rangle_0^2 &= \langle \phi, f\rangle_0^2 + \int_0^t 2\langle u_r^\epsilon, \phi\rangle_0 \left\langle Lu_r^\epsilon - \frac{\gamma}{2} u_r^\epsilon T_\epsilon v_r^{\epsilon,\epsilon}, \phi \right\rangle_0 dr \\ &\quad + \int_0^t \int_U 2\langle u_r^\epsilon, \phi\rangle_0 \langle h(y,\cdot)\nabla u_r^\epsilon, \phi\rangle_0 W(drdy) \\ &\quad + \int_0^t \int_U \langle h(y,\cdot)\nabla u_r^\epsilon, \phi\rangle_0^2 \mu(dy) dr. \end{aligned}$$

Adding over ϕ in a CONS of H_0, we obtain

$$\begin{aligned} \|u_t^\epsilon\|_0^2 &= \|f\|_0^2 + \int_0^t 2\left\langle u_r^\epsilon,\ Lu_r^\epsilon - \frac{\gamma}{2} u_r^\epsilon T_\epsilon v_r^{\epsilon,\epsilon} \right\rangle_0 dr \\ &\quad + \int_0^t \int_U 2\langle u_r^\epsilon, h(y,\cdot)\nabla u_r^\epsilon\rangle_0 W(drdy) \\ &\quad + \int_0^t \int_U \|h(y,\cdot)\nabla u_r^\epsilon\|_0^2 \mu(dy) dr. \end{aligned}$$

Applying Itô's formula again, we have

$$\|u_t^\epsilon\|_0^4 = \|f\|_0^4 + \int_0^t 4\|u_r^\epsilon\|_0^2 \left\langle u_r^\epsilon,\ Lu_r^\epsilon - \frac{\gamma}{2} u_r^\epsilon T_\epsilon v_r^{\epsilon,\epsilon} \right\rangle_0 dr$$
$$+ \int_0^t \int_U 4\|u_r^\epsilon\|_0^2 \langle u_r^\epsilon, h(y,\cdot)\nabla u_r^\epsilon\rangle_0 W(drdy)$$
$$+ \int_0^t \int_U 2\|u_r^\epsilon\|_0^2 \|h(y,\cdot)\nabla u_r^\epsilon\|_0^2 \mu(dy)dr$$
$$+ \int_0^t \int_U 4 \langle u_r^\epsilon, h(y,\cdot)\nabla u_r^\epsilon\rangle_0^2 \mu(dy)dr. \tag{5.2.9}$$

We remark that the only coefficient in (5.2.9) which depends on ϵ is $-(T_\epsilon v_r^{\epsilon,\epsilon})$, and since this term is negative, it can be discarded. The other terms in (5.2.9) can be estimated as follows: By (3.1.7) and (3.1.10) (with $\delta = 0$ there), we have

$$\langle u_r^\epsilon, Lu_r^\epsilon\rangle_0 + \int_U \frac{1}{2}\|h(y,\cdot)\nabla u_r^\epsilon\|_0^2 \mu(dy) \le K_2\|u_r^\epsilon\|_0^2 \tag{5.2.10}$$

and

$$\int_U \langle u_r^\epsilon, h(y,\cdot)\nabla u_r^\epsilon\rangle_0^2 \mu(dy) \le K_3\|u_r^\epsilon\|_0^4. \tag{5.2.11}$$

Therefore,

$$\|u_t^\epsilon\|_0^4 \le \|f\|_0^4 + K_4 \int_0^t \|u_r^\epsilon\|_0^4 dr$$
$$+ \int_0^t \int_U 4\|u_r^\epsilon\|_0^2 \langle u_r^\epsilon, h(y,\cdot)\nabla u_r^\epsilon\rangle_0 W(drdy).$$

By Burkholder-Davis-Gundy inequality and (5.2.10, 5.2.11), we then get

$$\mathbb{E}\sup_{s\le t}\|u_s^\epsilon\|_0^4 \le \|f\|_0^4 + K_4 \int_0^t \|u_r^\epsilon\|_0^4 dr$$
$$+K_5\mathbb{E}\left(\int_0^t \int_U \|u_r^\epsilon\|_0^4 \langle u_r^\epsilon, h(y,\cdot)\nabla u_r^\epsilon\rangle_0^2 \mu(dy)dr\right)^{1/2}$$
$$\le \|f\|_0^4 + K_4 \int_0^t \|u_r^\epsilon\|_0^4 dr$$
$$+K_6\mathbb{E}\left(\sup_{s\le t}\|u_s^\epsilon\|_0^2 \left(\int_0^t \|u_r^\epsilon\|_0^4 dr\right)^{1/2}\right)$$
$$\le \|f\|_0^4 + K_7 \int_0^t \|u_r^\epsilon\|_0^4 dr + \frac{1}{2}\mathbb{E}\sup_{s\le t}\|u_s^\epsilon\|_0^4.$$

Therefore,

$$\mathbb{E} \sup_{s \leq t} \|u_s^\epsilon\|_0^4 \leq 2\|f\|_0^4 + 2K_7 \int_0^t \mathbb{E}\|u_r^\epsilon\|_0^4 dr. \tag{5.2.12}$$

Gronwall's inequality implies that

$$\mathbb{E} \sup_{0 \leq t \leq T} \|u_t^\epsilon\|_0^4 \leq K_8. \tag{5.2.13}$$

Let $z_r^\epsilon = \nabla u_r^\epsilon$. Note that

$$u_r^\epsilon \nabla \left(T_\epsilon v_r^{\epsilon,\epsilon}\right) = u_r^\epsilon \lambda_\epsilon(u_r^\epsilon) \nabla \left(T_\epsilon v_r^\epsilon\right) = u^{\epsilon,\epsilon} T_\epsilon z_r^\epsilon,$$

where we used the identity of u^ϵ and v^ϵ, as well as the exchangeability of operators T_ϵ and ∇.

Taking the derivative on both sides of (5.2.8), we get

$$\begin{aligned} z_t^\epsilon(x) = f'(x) &+ \int_0^t \left(L^\epsilon z_r^\epsilon(x) - \frac{\gamma}{2} u_r^{\epsilon,\epsilon}(x) T_\epsilon z_r^\epsilon(x)\right) dr \\ &+ \int_0^t \int_U \left(\nabla h(y,x) z_r^\epsilon(x) + h(y,x) \nabla z_r^\epsilon(x)\right) W(drdy), \end{aligned}$$

where

$$L^\epsilon f = \frac{1}{2} a f'' + b f' + c^\epsilon f,$$

and

$$b = \frac{1}{2} a', \qquad c^\epsilon = -\frac{\gamma}{2} T_\epsilon u^{\epsilon,\epsilon}.$$

Observe that

$$u^{\epsilon,\epsilon}, \ c^\epsilon, \ \text{ and } \int_U |\nabla h(y,\cdot)|^2 \mu(dy)$$

are all bounded, and for $z \in H_0$,

$$\begin{aligned} |\langle z, T_\epsilon z\rangle_0| &\leq \int_{\mathbb{R}} \int_{\mathbb{R}} p_\epsilon(x-y) |z(x)||z(y)| dx dy \\ &\leq \int_{\mathbb{R}} \int_{\mathbb{R}} p_\epsilon(x-y) \frac{1}{2} \left(|z(x)|^2 + |z(y)|^2\right) dx dy \\ &= \|z\|_0^2. \end{aligned}$$

Similar to arguments leading to (5.2.13), we arrive at

$$\mathbb{E} \sup_{0 \leq t \leq T} \|z_t^\epsilon\|_0^4 \leq K_1. \tag{5.2.14}$$

The conclusion of the lemma then follows from (5.2.13) and (5.2.14). □

5.3 Existence and uniqueness

In this section, we prove the first part of Theorem 5.1.2.

Proof. For ϵ, $\eta > 0$, let

$$z_t(x) \equiv v_t^{\epsilon,\eta}(x) \equiv v_t^\epsilon(x) - v_t^\eta(x),$$

where $v_t^\eta(x)$ is the solution of SPDE (5.2.1) with ϵ replaced by η. Taking the difference, we obtain

$$\begin{aligned} z_t(x) = & \int_0^t \left(Lz_r(x) - \frac{\gamma}{2} z_r(x) T_\epsilon v_r^{\epsilon,\epsilon}(x) \right) dr \\ & + \int_0^t \int_U h(y,x) \nabla z_r(x) W(drdy) \\ & - \int_0^t \frac{\gamma}{2} v_r^\eta(x) \left(T_\epsilon v_r^{\epsilon,\epsilon}(x) - T_\eta v_r^{\eta,\eta}(x) \right) dr. \end{aligned}$$

Note that

$$\begin{aligned} T_\epsilon v_r^{\epsilon,\epsilon} - T_\eta v_r^{\eta,\eta} = & \lambda_\epsilon(v^\epsilon) T_\epsilon z_r + \left(\lambda_\epsilon(v_r^\epsilon) - \lambda_\eta(v_r^\eta) \right) T_\epsilon v_r^\eta \\ & + \lambda_\eta(v^\eta) \left(T_\epsilon v_r^\eta - T_\eta v_r^\eta \right). \end{aligned}$$

Similar to (5.2.12), we have

$$\begin{aligned} & \mathbb{E} \sup_{0 \le s \le t} \|z_s\|_0^4 \\ \le & K_1 \int_0^t \mathbb{E}\|z_r\|_0^4 dr + K_2 \mathbb{E} \int_0^t |\lambda_\epsilon(v_r^\epsilon) - \lambda_\eta(v_r^\eta)|^4 dr \\ & + 3\|f\|_\infty^4 \mathbb{E} \int_0^t \left(\int_{\mathbb{R}} |T_\epsilon v_r^\eta(x) - T_\eta v_r^\eta(x)|^2 dx \right)^2 dr. \qquad (5.3.1) \end{aligned}$$

As

$$\begin{aligned} & T_\epsilon v_r^\eta(x) - T_\eta v_r^\eta(x) \\ = & \int_{\mathbb{R}} p_1(a) \left(v_r^\eta(x + \epsilon a) - v_r^\eta(x + \eta a) \right) da \\ = & \int_{\mathbb{R}} \int_0^1 \nabla v_r^\eta(x + (\theta\sqrt{\epsilon} + (1-\theta)\sqrt{\eta})a)(\sqrt{\epsilon} - \sqrt{\eta}) a d\theta p_1(a) da, \end{aligned}$$

we obtain, when $\epsilon,\ \eta \to 0$,

$$\begin{aligned}
&\int_{\mathbb{R}} |T_\epsilon v_r^\eta(x) - T_\eta v_r^\eta(x)|^2 dx \\
&\leq \int_{\mathbb{R}}\int_{\mathbb{R}}\int_0^1 \left|\nabla v_r^\eta(x+(\theta\sqrt{\epsilon}+(1-\theta)\sqrt{\eta})a)\right|^2 \left(\sqrt{\epsilon}-\sqrt{\eta}\right)^2 a^2 d\theta p_1(a) da dx \\
&= \int_{\mathbb{R}}\int_0^1 \|\nabla v_r^\eta\|_0^2 \left(\sqrt{\epsilon}-\sqrt{\eta}\right)^2 a^2 d\theta p_1(a) da \\
&= \|\nabla v_r^\eta\|_0^2 (\sqrt{\epsilon}-\sqrt{\eta})^2 \\
&\to 0.
\end{aligned} \tag{5.3.2}$$

By Corollary 5.2.3 and the dominated convergence theorem, we have

$$\mathbb{E}\int_0^t |\lambda_\epsilon(v_r^\epsilon) - \lambda_\eta(v_r^\eta)|^4 dr \to 0. \tag{5.3.3}$$

It follows from Gronwall's inequality, (5.3.1), (5.3.2) and (5.3.3) that

$$\mathbb{E}\sup_{0\leq t\leq T} \|v_t^\epsilon - v_t^\eta\|_0^4 \to 0 \qquad \text{as } \epsilon,\ \eta \to 0.$$

Hence, there exists a stochastic process (v_t) taking values in $C([0,T],H_0)$ a.s. such that

$$\mathbb{E}\sup_{0\leq t\leq T} \|v_t^\epsilon - v_t\|_0^4 \to 0 \qquad \text{as } \epsilon \to 0.$$

Recall that

$$\begin{aligned}
\langle v_t^\epsilon, \phi\rangle_0 = \langle \phi, f\rangle_0 &+ \int_0^t \left\langle v_r^\epsilon, L^*\phi - \frac{\gamma}{2}\phi T_\epsilon v_r^{\epsilon,\epsilon}\right\rangle_0 dr \\
&- \int_0^t \int_U \langle v_r^\epsilon, \nabla(h(y,\cdot)\phi)\rangle_0 W(drdy).
\end{aligned}$$

We consider the limit of the nonlinear term only, since the other terms clearly converge to the counterpart with v^ϵ replaced by v. Note that

$$\begin{aligned}
&\mathbb{E}\left|\int_0^t\int_{\mathbb{R}} \phi(x) v_r^\epsilon(x) T_\epsilon v_r^{\epsilon,\epsilon}(x) dx dr - \int_0^t\int_{\mathbb{R}} v_r(x)^2\phi(x) dx dr\right| \\
&\leq \mathbb{E}\int_0^t\int_{\mathbb{R}} |T_\epsilon(v_r^{\epsilon,\epsilon} - v_r)(x)| |v_r^\epsilon(x)| |\phi(x)| dx dr \\
&\quad + \mathbb{E}\int_0^t\int_{\mathbb{R}} |T_\epsilon v_r(x) - v_r(x)| |v_r^\epsilon(x)| |\phi(x)| dx dr \\
&\quad + \mathbb{E}\int_0^t\int_{\mathbb{R}} |v_r^\epsilon(x) - v_r(x)| |v_r(x)| |\phi(x)| dx dr \\
&\to 0.
\end{aligned}$$

It is then easy to show that v_t solves (5.1.4).

To prove the uniqueness for the solution of equation (5.1.4), we assume that v_t and $\tilde{v}_t$ are two solution to (5.1.4). Similar to (5.3.1), we have

$$\mathbb{E}\sup_{s\le t}\|v_t-\tilde{v}_t\|_0^4\le K_1\int_0^t\mathbb{E}\|v_r-\tilde{v}_r\|_0^4dr. \tag{5.3.4}$$

Uniqueness then follows from Gronwall's inequality. □

The next result is an immediate consequence of (5.2.7) and the convergence of v^ϵ to v.

Corollary 5.3.1.

$$\mathbb{E}\sup_{0\le t\le T}\|\nabla v_t\|_0^4<\infty.$$

Proof. Let $\{\phi_i\}\subset H_1$ be a CONS of H_0. Then,

$$\begin{aligned}\mathbb{E}\sup_{0\le t\le T}\|\nabla v_t\|_0^4&=\mathbb{E}\left(\sup_{0\le t\le T}\sum_i\langle\nabla v_t,\phi_i\rangle^2\right)^2\\&=\mathbb{E}\left(\sup_{0\le t\le T}\sum_i\langle v_t,\phi_i'\rangle^2\right)^2\\&=\mathbb{E}\left(\sup_{0\le t\le T}\sum_i\lim_{\epsilon\to0}\langle v_t^\epsilon,\phi_i'\rangle^2\right)^2\\&\le\liminf_{\epsilon\to0}\mathbb{E}\left(\sup_{0\le t\le T}\sum_i\langle v_t^\epsilon,\phi_i'\rangle^2\right)^2\\&=\liminf_{\epsilon\to0}\mathbb{E}\sup_{0\le t\le T}\|\nabla v_t^\epsilon\|_0^4\\&\le K_1,\end{aligned}$$

where K_1 is the constant (independent of ϵ) given in (5.2.7). □

Finally, we verify (ii) of Theorem 5.1.2.

Proof. Let v_t be the solution to (5.1.4) and $\nu_t(dx)=v_t(x)dx$. Let (ξ_t^i,m_t^i) be given by (5.1.5, 5.1.6), and denote the process given by the right hand side of (5.1.7) by $\tilde{\nu}_t$. Now we only need to verify that $\tilde{\nu}_t$ coincides with ν_t. Applying Itô's formula to $m_t^i\phi(\xi_t^i)$, it is easy to show that

$$\begin{aligned}\langle\tilde{\nu}_t,\phi\rangle=\langle f,\phi\rangle&+\int_0^t\left\langle\tilde{\nu}_r,L^*\phi-\frac{\gamma}{2}v_r\phi\right\rangle dr\\&-\int_0^t\int_U\langle\tilde{\nu}_r,\nabla(h(y,\cdot)\phi)\rangle W(drdy).\end{aligned} \tag{5.3.5}$$

By (5.1.4), we see that (5.3.5) also holds with $\tilde{\nu}_t$ replaced by ν_t. By the uniqueness of the solution to linear SPDE (see Theorem 3.1.7), we have $\nu_t = \tilde{\nu}_t$ and hence, ν_t has the particle representation given in Theorem 5.1.2. □

5.4 Conditional log-Laplace transform

In this section, we prove that the solution of SPDE (5.1.3) is the logarithm of the conditional Laplace transform of an SPRE.

The idea is to approximate the SPRE and the solution to SPDE (5.1.3) simultaneously. More specifically, we divide the time line $\mathbb{R}_+$ into small subintervals of length $\epsilon > 0$ each. In the first time period, the approximate process is a classical superprocess (i.e. $h = 0$) with branching rate 2γ and initial ν; while in the second time period, it is the solution to a linear SPDE (i.e. $\gamma = 0$) with initial coincides with the terminal of the superprocess in the first period. This procedure is continued for all time periods, i.e., the classical superprocess and linear SPDE occur alternatively with initial of the proceeding process equal to the terminal of the preceding one.

For the classical superprocess, the Laplace transform, and hence the conditional Laplace transform, is represented by the solution to a nonlinear PDE; For the solution to a linear SPDE, we apply the duality (3.2.18) established in Section 3.2 to obtain the logarithm of the conditional Laplace transform by another linear SPDE. Thus, the conditional Laplace transform of the approximating process is given alternatively by nonlinear PDEs and linear SPDEs. The desired conditional Laplace transform for the SPRE in the whole time line is then obtained by taking $\epsilon \to 0$.

Now, we define the aforementioned approximation more precisely. In the intervals $[2i\epsilon, (2i+1)\epsilon]$, $i = 0, 1, 2, \cdots$, X^ϵ is a superprocess (with deterministic environment) with initial $X^\epsilon_{2i\epsilon}$, i.e., $\forall\ \phi \in C^2_b(\mathbb{R})$ and $t \in [2i\epsilon, (2i+1)\epsilon]$,

$$M^\epsilon_t(\phi) \equiv \langle X^\epsilon_t, \phi\rangle - \langle X^\epsilon_{2i\epsilon}, \phi\rangle - \int_{2i\epsilon}^t \langle X^\epsilon_s, L\phi\rangle\, ds$$

is a continuous martingale with quadratic variation process

$$\langle M^\epsilon(\phi)\rangle_t = 2\gamma \int_{2i\epsilon}^t \left\langle X^\epsilon_s, \phi^2\right\rangle ds;$$

and in the intervals $[(2i+1)\epsilon, 2(i+1)\epsilon]$, it is the solution to the following linear SPDE: $\forall\ \phi \in C_b^2(\mathbb{R})$ and $t \in [(2i+1)\epsilon, 2(i+1)\epsilon]$,

$$\langle X_t^\epsilon, \phi\rangle = \left\langle X_{(2i+1)\epsilon}^\epsilon, \phi\right\rangle + \int_{(2i+1)\epsilon}^t \langle X_s^\epsilon, L\phi\rangle\, ds$$
$$+\sqrt{2}\int_{(2i+1)\epsilon}^t \int_U \langle X_s^\epsilon, h(y,\cdot)\phi'\rangle\, W(dsdy).$$

Let $\{\phi^j\}$ be a CONS of $L^2(U,\mu)$. We define a sequence of stochastic processes

$$W_t^{j,\epsilon} = \sqrt{2}\int_0^t \int_U 1_{A^c}(s)\phi^j(y)W(dsdy), \qquad j = 1, 2, \cdots, \tag{5.4.1}$$

where

$$A = \{s :\ 2i\epsilon \le s \le (2i+1)\epsilon,\ i = 0, 1, 2, \cdots\}.$$

Denote the $\mathbb{R}^\infty$-valued process

$$W^\epsilon = \left(W^{1,\epsilon}, W^{2,\epsilon}, \cdots\right).$$

It is easy to see that $\{X^\epsilon, W^\epsilon\}$ is a solution to the following approximate martingale problem (AMP): $W^{j,\epsilon}$, $j = 1, 2, \cdots$ are independent continuous martingales with

$$\left\langle W^{j,\epsilon}\right\rangle_t = \int_0^t \int_U 1_{A^c}(s)\phi^j(y)^2\mu(dy)ds,$$

and for any $\phi,\ \psi \in C_b^2(\mathbb{R})$,

$$\langle X_t^\epsilon, \phi\rangle = \langle \mu, \phi\rangle + \int_0^t \langle X_s^\epsilon, L\phi\rangle\, ds + M_t^{1,\epsilon}(\phi) + M_t^{2,\epsilon}(\phi)$$

where $M_t^{1,\epsilon}(\phi)$ and $M_t^{2,\epsilon}(\psi)$ are uncorrelated martingales satisfying

$$\left\langle M^{1,\epsilon}(\phi)\right\rangle_t = 2\gamma\int_0^t \left\langle X_s^\epsilon, \phi^2\right\rangle 1_A(s)ds,$$

and

$$\left\langle M^{2,\epsilon}(\phi)\right\rangle_t = 2\int_0^t \int_U \langle X_s^\epsilon, h(y,\cdot)\phi'\rangle^2\, 1_{A^c}(s)\mu(dy)ds;$$

and for $j = 1,\ 2,\ \cdots$,

$$\left\langle M^{1,\epsilon}(\phi), W^{j,\epsilon}\right\rangle_t = 0,$$

and

$$\left\langle M^{2,\epsilon}(\phi), W^{j,\epsilon}\right\rangle_t = 2\int_0^t \int_U \langle X_s^\epsilon, h(y,\cdot)\phi'\rangle\, \phi^j(y)1_{A^c}(s)\mu(dy)ds.$$

In the next couple of lemmas, we prove the tightness of $\{X^\epsilon\}$ and $\{W^\epsilon\}$, respectively.

Lemma 5.4.1. *The family $\{X^\epsilon\}$ is tight in $C(\mathbb{R}_+, \mathcal{M}_F(\mathbb{R}))$.*

Proof. Notice that

$$\langle X_t^\epsilon, 1\rangle = \langle \nu, 1\rangle + M_t^{1,\epsilon}(1).$$

Applying Burkholder-Davis-Gundy inequality, we get

$$\begin{aligned}\mathbb{E}\sup_{s\le t}\langle X_t^\epsilon, 1\rangle^4 &\le 4\langle \nu, 1\rangle^4 + 4\left(\frac{4}{4-1}\right)^4 \mathbb{E}\left(2\gamma\int_0^t \langle X_s^\epsilon, 1\rangle 1_A(s)ds\right)^2\\ &\le K_1 + K_2\int_0^t \mathbb{E}\langle X_s^\epsilon, 1\rangle^4 ds.\end{aligned}$$

It follows from Gronwall's inequality that

$$\mathbb{E}\sup_{s\le t}\langle X_t^\epsilon, 1\rangle^4 \le K_3. \tag{5.4.2}$$

Next, for $0 < s < t$ and $\phi \in C_b^2(\mathbb{R})$, we have

$$\begin{aligned}&\mathbb{E}\left|\langle X_t^\epsilon, \phi\rangle - \langle X_s^\epsilon, \phi\rangle\right|^4\\ &= 8\mathbb{E}\left|\int_s^t \langle X_r^\epsilon, L\phi\rangle\, dr\right|^4 + 8\mathbb{E}\left|\int_s^t\int_U 1_A(r)\langle X_r^\epsilon, h(y,\cdot)\phi'\rangle\, W(drdy)\right|^4\\ &\quad + \mathbb{E}\left|\int_s^t 1_{A^c}(r)dM_r^{1,\epsilon}(\phi)\right|^4\\ &\le 8\mathbb{E}\left|\int_s^t \langle X_r^\epsilon, L\phi\rangle\, dr\right|^4 + 8\left(\frac{4}{4-1}\right)^4\mathbb{E}\left(2\gamma\int_s^t \langle X_s^\epsilon, \phi^2\rangle 1_A(s)ds\right)^2\\ &\quad + 8\left(\frac{4}{4-1}\right)^4\mathbb{E}\left(\int_s^t\int_U 1_A(r)\langle X_r^\epsilon, h(y,\cdot)\phi'\rangle^2 \mu(dy)dr\right)^2\\ &\le K_4|t-s|^2.\end{aligned} \tag{5.4.3}$$

Note that the constant K_4 in (5.4.3) depends on $\|\phi\|_{2,\infty}$ only, where

$$\|\phi\|_{2,\infty} = \sum_{\beta=0}^2 \sup_x |\phi^{(\beta)}(x)|.$$

We can take a sequence $\{f_n\}$ in $C_b^2(\mathbb{R})$ such that $\|f_n\|_{2,\infty} \le 1$ for all $n \ge 1$. The weak topology of $\mathcal{M}_F(\bar{\mathbb{R}})$ is given by the metric d defined by

$$d(\mu,\nu) \equiv \sum_{n=1}^\infty 2^{-n}\left(|\langle \mu-\nu, f_n\rangle| \wedge 1\right),$$

where $\bar{\mathbb{R}}$ is the compactification of $\mathbb{R}$. By (5.4.3), it is easy to show that

$$\mathbb{E}d(X_t^\epsilon, X_s^\epsilon)^4 \le K_4|t-s|^2. \tag{5.4.4}$$

The tightness of X^ϵ in $C(\mathbb{R}_+, \mathcal{M}_F(\bar{\mathbb{R}}))$ follows from (5.4.2) and (5.4.4) (cf. Corollary 16.9 in Kallenberg (2002), p. 313). $\square$

If $R\to\infty$, then

$$\mathbb{E}\left(X_t^\epsilon(\phi_R)\right)=X_0(T_t^{(1)}\phi_R)\to 0,$$

where

$$\phi_R(x)=\begin{cases}0 & \text{for } |x|\le R,\\ 1 & \text{for } |x|\ge R+1\\ \text{linear connection} & \text{for } R<|x|<R+1.\end{cases}$$

Then for any $t\ge 0$, we have $X_t(\{\infty\})=0$. By the continuity of X, we obtain that almost surely, $X_t(\{\infty\})=0$ for all $t\ge 0$. Thus, X^ϵ is tight in $C(\mathbb{R}_+,\mathcal{M}_F(\mathbb{R}))$. □

To study the tightness of W^ϵ, we need to define a space for which W^ϵ take its values. For any $g\in\mathbb{R}^\infty$, we define norms $\|g\|_i$, $i=1,\ 2$, by

$$\|g\|_i^2\equiv\sum_{j=1}^\infty j^{-2i}g_j^2.$$

Let $\mathbb{L}_i$ be the completion of ℓ^2 with respect to $\|\cdot\|_i$, where

$$\ell^2=\left\{g\in\mathbb{R}^\infty:\ \sum_{j=1}^\infty g_j^2<\infty\right\}.$$

Then $\ell^2\subset\mathbb{L}_1\subset\mathbb{L}_2$ and the injections are compact.

Lemma 5.4.2. *$\{W^\epsilon\}$ is tight in $C(\mathbb{R}_+,\mathbb{L}_2)$.*

Proof. Note that

$$\begin{aligned}\mathbb{E}\sup_{r\le t}\|W_r^\epsilon\|_1^2&=\mathbb{E}\sup_{r\le t}\sum_{j=1}^\infty j^{-2}\left(\sqrt{2}\int_0^r\int_U 1_{A^c}(s)\phi^j(y)W(dsdy)\right)^2\\ &\le 2\sum_{j=1}^\infty j^{-2}\mathbb{E}\sup_{r\le t}\left(\sqrt{2}\int_0^r\int_U 1_{A^c}(s)\phi^j(y)W(dsdy)\right)^2\\ &\le 2\sum_{j=1}^\infty j^{-2}4\mathbb{E}\int_0^t 1_{A^c}(s)ds\\ &\le 16t.\end{aligned}$$

Since the injection from $\mathbb{L}_1$ to $\mathbb{L}_2$ is compact, $\{x\in\mathbb{L}_2:\ \|x\|_1\le K\}$ is compact in $\mathbb{L}_2$. Thus, $\{W^\epsilon\}$ satisfies the compact containment condition in $\mathbb{L}_2$.

Similarly, we can prove that, for any $s<t$ and $\epsilon>0$,

$$\mathbb{E}\left(\|W_t^\epsilon-W_s^\epsilon\|_2^4\right)\le\mathbb{E}\left(\|W_t^\epsilon-W_s^\epsilon\|_1^4\right)\le K_1|t-s|^2.$$

The tightness of $\{W^\epsilon\}$ in $C(\mathbb{R}_+,\mathbb{L}_2)$ then follows easily. □

Lemma 5.4.3. *Let (X^0, W^0) be any limit point of $\{(X^\epsilon, W^\epsilon)\}$. Then (X^0, W^0) satisfies the following joint martingale problem (JMP): $W^0 = \{W^{j,0} : j \in \mathbb{N}\}$ is an i.i.d. sequence of Brownian motions, and $\forall \phi \in C_b^2(\mathbb{R})$,*

$$M_t(\phi) \equiv \langle X_t^0, \phi \rangle - \langle \nu, \phi \rangle - \int_0^t \langle X_s^0, L\phi \rangle \, ds$$

is a continuous martingale with quadratic variation processes

$$\langle M(\phi) \rangle_t = \int_0^t \langle X_s^0, \gamma\phi^2 \rangle \, ds + \int_0^t \int_U \langle X_s^0, h(y, \cdot)\phi' \rangle^2 \mu(dy) ds$$

and for $j = 1, 2, \cdots$,

$$\langle M(\phi), W^{j,0} \rangle_t = \int_0^t \int_U \langle X_s^0, h(y, \cdot)\phi' \rangle \phi^j(y) \mu(dy) ds.$$

Moreover, X^0 satisfies MP (2.1.6, 2.1.7).

Proof. This in immediate from the previous lemmas and the convergence of all the terms in AMP to the corresponding terms in JMP. □

Now we define an approximation $v_{s,t}^\epsilon$ of $v_{s,t}$ in two cases. Note that in our construction, the process $v_{s,t}^\epsilon$ will be independent of X^ϵ *conditionally on* W^ϵ. First, we suppose that $2k\epsilon \le t < (2k+1)\epsilon$. Then for $2k\epsilon \le s \le t$, we define $v_{s,t}^\epsilon$ by the following nonlinear PDE:

$$v_{s,t}^\epsilon = \phi + \int_s^t \left(L v_{r,t}^\epsilon - \frac{\gamma}{2} (v_{r,t}^\epsilon)^2 \right) dr,$$

and for $(2k-1)\epsilon \le s \le 2k\epsilon$, let $v_{s,t}^\epsilon$ be the solution to the following backward linear SPDE:

$$v_{s,t}^\epsilon = v_{2k\epsilon,t}^\epsilon + \int_s^{2k\epsilon} L v_{r,t}^\epsilon dr + \int_s^{2k\epsilon} \int_U \nabla v_{r,t}^\epsilon h(y, \cdot) W^\epsilon(\hat{d}r dy).$$

The definition continues in this pattern. For the case of $(2k+1)\epsilon \le t < 2(k+1)\epsilon$, the definition is modified in an obvious manner.

Since the behavior of the processes X_s^ϵ and $v_{s,t}^\epsilon$ does not depend on W^ϵ we get

$$\mathbb{E}^{W^\epsilon} \left(e^{-\langle X_t^\epsilon, \phi \rangle} | X_{2k\epsilon} \right) = e^{-\left\langle X_{2k\epsilon}^\epsilon, v_{2k\epsilon,t}^\epsilon \right\rangle}.$$

Hence,

$$\mathbb{E}^{W^\epsilon} e^{-\langle X_t^\epsilon, \phi \rangle} = \mathbb{E}^{W^\epsilon} e^{-\left\langle X_{2k\epsilon}^\epsilon, v_{2k\epsilon,t}^\epsilon \right\rangle}.$$

Using Theorem 3.2.5, we have

$$\left\langle X^\epsilon_{2k\epsilon}, v^\epsilon_{2k\epsilon,t}\right\rangle = \left\langle X^\epsilon_{(2k-1)\epsilon}, v^\epsilon_{(2k-1)\epsilon,t}\right\rangle, \ P-\text{a.s.},$$

therefore,

$$\mathbb{E}^{W^\epsilon} e^{-\langle X^\epsilon_t,\phi\rangle} = \mathbb{E}^{W^\epsilon} e^{-\left\langle X^\epsilon_{(2k-1)\epsilon}, v^\epsilon_{(2k-1)\epsilon,t}\right\rangle}.$$

By continuing this pattern, we obtain

$$\mathbb{E}^{W^\epsilon} e^{-\langle X^\epsilon_t,\phi\rangle} = e^{-\left\langle \nu, v^\epsilon_{0,t}\right\rangle}. \tag{5.4.5}$$

Note again that in our construction the process $v^\epsilon_{s,t}$ is independent of X^ϵ conditionally on W^ϵ.

Lemma 5.4.4. *We endow H_0 with weak topology. Then for any $t>0$, $\{v^\epsilon_{\cdot,t}\}$ is tight in $C([0,t],H_0)$.*

Proof. For the simplicity of presentation, we will consider the forward version of the equations. Also, we assume that $t=k'\epsilon$, and will consider the case of $k'=2k$ only, since the other case can be treated similarly. Let $\mathbb{W}^\epsilon_s = W^\epsilon_{t-s} - W^\epsilon_t$. Then, for $2i\epsilon \le s \le (2i+1)\epsilon$, $0\le i<k$,

$$v^\epsilon_s = v^\epsilon_{2i\epsilon} + \int_{2i\epsilon}^s Lv^\epsilon_r dr + \int_{2i\epsilon}^s \int_U \nabla v^\epsilon_r h(y,\cdot)\mathbb{W}^\epsilon(drdy), \tag{5.4.6}$$

and for $(2i+1)\epsilon \le s \le 2(i+1)\epsilon$, $0\le i<k$,

$$v^\epsilon_s = v^\epsilon_{(2i+1)\epsilon} + \int_{(2i+1)\epsilon}^s \left(Lv^\epsilon_r - \frac{\gamma}{2}(v^\epsilon_r)^2\right) dr. \tag{5.4.7}$$

It is easy to show that the solution of (5.4.6) is an increasing functional of the initial condition $v^\epsilon_{2i\epsilon}$; and the solution of (5.4.7) is less than ϕ^ϵ_s given by

$$\phi^\epsilon_s = \phi^\epsilon_{(2i+1)\epsilon} + \int_{(2i+1)\epsilon}^s L\phi^\epsilon_r dr, \qquad (2i+1)\epsilon \le s \le 2(i+1)\epsilon$$

provided that $\phi^\epsilon_{(2i+1)\epsilon} \ge v^\epsilon_{(2i+1)\epsilon}$. For $2i\epsilon \le s\le (2i+1)\epsilon$, we define v^ϵ_s by (5.4.6) with $v^\epsilon_{2i\epsilon}$ replaced by $\phi^\epsilon_{2i\epsilon}$. Then, $v^\epsilon_s \le \phi^\epsilon_s$ for all $s\in[0,t]$.

Note that

$$\phi^\epsilon_s = \phi + \int_0^s L\phi^\epsilon_r dr + \int_0^s 1_{A^c}(r)\phi^\epsilon_r \left\langle g, d\mathbb{W}^\epsilon_r\right\rangle_{\mathbb{H}}.$$

Now we use the same arguments as those in Section 3.1. We let $Z^{\delta,\epsilon}_s = T_\delta \phi^\epsilon_s$ and apply Itô's formula to obtain an expression for $\|Z^{\delta,\epsilon}_s\|_0^2$ and then, take $\delta\to 0$ to derive,

$$\mathbb{E}\sup_{r\le s}\|\phi^\epsilon_r\|_0^{2p} \le K_1 + K_2\int_0^s \mathbb{E}\|\phi^\epsilon_r\|_0^{2p} dr. \tag{5.4.8}$$

By Gronwall's inequality, we have

$$\mathbb{E}\sup_{r\le s}\|v_r^\epsilon\|_0^{2p} \le \mathbb{E}\sup_{r\le s}\|\phi_r^\epsilon\|_0^{2p} \le K_1 e^{K_2 t}. \tag{5.4.9}$$

Since

$$v_t^\epsilon = \phi + \int_0^t Lv_r^\epsilon dr + \int_0^t \int_U 1_{A^c}(r)\nabla v_r^\epsilon h(y,\cdot)\mathbb{W}^\epsilon(drdy)$$
$$-\gamma\int_0^t 1_A(r)(v_r^\epsilon)^2 dr,$$

then for any $f \in H_0 \cap C_b^2(\mathbb{R})$, we have

$$\begin{aligned}\mathbb{E}\left|\langle v_t^\epsilon - v_s^\epsilon, f\rangle\right|_0^{2p} \le\; & 3^{2p-1}\mathbb{E}\left|\int_s^t \langle v_r^\epsilon, Lf\rangle\, dr\right|^{2p}\\ &+ 3^{2p-1}\mathbb{E}\left|\int_s^t\int_U 1_{A^c}(r)\,\langle v_r^\epsilon h(y,\cdot), f'\rangle\,\mathbb{W}^\epsilon(drdy)\right|^{2p}\\ &+ 3^{2p-1}\mathbb{E}\left|\gamma\int_s^t 1_A(r)\int v_r^\epsilon(x)^2|f(x)|dxdr\right|^{2p}\\ \le\; & K_3|t-s|^p.\end{aligned} \tag{5.4.10}$$

The tightness of $\{v^\epsilon\}$ then follows from (5.4.9) and (5.4.10) with $p > 1$. □

Corollary 5.4.5. *Let v^ϵ be a solution to (5.4.6)-(5.4.7). Then, $\{v^\epsilon\}$ is tight in $C(\mathbb{R}_+, H_0)$.*

Proof. This is immediate from the previous lemma. □

Lemma 5.4.6. *Suppose that $(v^0, \mathbb{W}^0)$ is a limit point of $(v^\epsilon, \mathbb{W}^\epsilon)$. Then*

$$v_s^0 = \phi + \int_0^s \left(Lv_r^0 - \frac{\gamma}{2}(v_r^0)^2\right)dr + \int_0^s\int_U \nabla(v_r^0 h(y,\cdot))\mathbb{W}^0(drdy). \tag{5.4.11}$$

Similarly, let $\{v_{\cdot,t}^0, W^0)$ be a limit point of $\{v_{\cdot,t}^\epsilon, W^\epsilon)$. Then

$$v_{s,t}^0 = \phi + \int_s^t \left(Lv_{r,t}^0 - \frac{\gamma}{2}(v_{r,t}^0)^2\right)dr + \int_s^t\int_U \nabla(v_{r,t}^0 h(y,\cdot))W^0(\hat{d}rdy).$$

Proof. Note that for any $f \in C_0^2(\mathbb{R})$,

$$N_t^\epsilon(f) \equiv \langle v_t^\epsilon, f\rangle - \langle\phi, f\rangle - \int_0^t \left(\langle v_r^\epsilon, Lf\rangle - 1_A(r)\left\langle\gamma(v_r^\epsilon)^2, f\right\rangle\right)dr$$

is a martingale with

$$\langle N^\epsilon(f)\rangle_t = 2\int_0^t\int_U 1_{A^c}(r)\,\langle v_r^\epsilon, f'h(y,\cdot)\rangle^2\,\mu(dy)dr$$

and for any $j \in \mathbb{N}$,

$$\left\langle N^\epsilon(f), \mathbb{W}^{j,\epsilon}\right\rangle_t = 2\int_0^t 1_{A^c}(r)\left\langle v_r^\epsilon, f'h(y,\cdot)\right\rangle \phi^j(y)\mu(dy)dr.$$

Passing to the limit, we see that

$$N_t(f) \equiv \left\langle v_t^0, f\right\rangle - \langle \phi, f\rangle - \int_0^t \left(\left\langle v_r^0, Lf\right\rangle - \left\langle \frac{\gamma}{2}(v_r^0)^2, f\right\rangle\right) dr$$

is a martingale with

$$\left\langle N^0(f)\right\rangle_t = \int_0^t \int_U \left\langle v_r^0, f'h(y,\cdot)\right\rangle^2 \mu(dy)dr$$

and for any $j,k \in \mathbb{N}$,

$$\left\langle \mathbb{W}^{j,0}, \mathbb{W}^{k,0}\right\rangle_t = \delta_{jk}t$$

and

$$\left\langle N^0(f), \mathbb{W}^{j,0}\right\rangle_t = \int_0^t \left\langle v_r^0, f'h(y,\cdot)\right\rangle \phi^j(y)\mu(dy)dr.$$

It is then easy to show that

$$\begin{aligned} N_t^0(f) &= \sum_{j=1}^\infty \int_0^t \int_U \left\langle v_r^0, f'h(y,\cdot)\right\rangle \phi^j(y)\mu(dy)d\mathbb{W}_r^{j,0} \\ &= \int_0^t \int_U \left\langle v_r^0, f'h(y,\cdot)\right\rangle \mathbb{W}^0(drdy), \end{aligned}$$

where

$$\mathbb{W}^0([0,t]\times B) = \sum_{j=1}^\infty \int_B \phi^j(y)\mu(dy)\mathbb{W}_t^{j,0}.$$

Thus, (5.4.11) holds. □

With the above preparation, we can prove the following theorem which demonstrates that the solution to (5.1.3) is the logarithm of the conditional Laplace transform of the SPRE.

Theorem 5.4.7. *The backward SPDE (5.1.3) has a pathwise unique non-negative solution $v_{s,t}(x)$. Moreover, there exists a triple (X^0, W^0, v^0) such that X and X^0 have the same law, and for any $\nu \in \mathcal{M}_F(\mathbb{R})$, we have*

$$\mathbb{E}^{W^0} \exp\left(-\left\langle X_t^0, \phi\right\rangle\right) = \exp\left(-\left\langle \nu, v_{0,t}^0\right\rangle\right). \tag{5.4.12}$$

Proof. Making use of Lemmas 5.4.1, 5.4.2, 5.4.3, 5.4.4 and 5.4.6, it follows from (5.4.5) that for a real valued continuous function, F, on $C([0,t],\mathbb{L}_2)$, we have

$$\begin{aligned} \mathbb{E}\left(\exp\left(-\left\langle \nu, v_{0,t}^0\right\rangle\right) F(W^0)\right) &= \lim_{\epsilon\to 0} \mathbb{E}\left(\exp\left(-\left\langle \nu, v_{0,t}^\epsilon\right\rangle\right) F(W^\epsilon)\right) \\ &= \lim_{\epsilon\to 0} \mathbb{E}\left(\exp\left(-\left\langle X_t^\epsilon, \phi\right\rangle\right) F(W^\epsilon)\right) \\ &= \mathbb{E}\left(\exp\left(-\left\langle X_t^0, \phi\right\rangle\right) F(W^0)\right). \end{aligned}$$

□

5.5 Historical remarks

The material of this chapter is based on the paper Xiong (2004a). Section 5.4 is adapted from a similar result of Mytnik and Xiong (2007). Applications of the log-Laplace equation can be found in Xiong (2004b) and Li, Wang and Xiong (2005).

Chapter 6

SPDEs for Density Fields of the Superprocesses in Random Environment

6.1 Introduction

In this chapter, we study an SPDE which is the counterpart of SPDE (1.4.3) for the superprocess in a random environment. Recall that κ was defined at the beginning of Section 3.3. We shall prove that the SPRE has density with respect to the Lebesgue measure and density random field $v_t(x)$ satisfies the following nonlinear SPDE:

$$v_t(x) = v(x) + \int_0^t \kappa \Delta v_s(x) ds - \int_0^t \int_U \nabla(h(y,x)v_s(x))W(dsdy) + \int_0^t \sqrt{\gamma v_s(x)} \frac{B(dsdx)}{dx}, \qquad (6.1.1)$$

where B is a white noise random measure on $\mathbb{R}_+ \times \mathbb{R}$ with intensity measure dx on $\mathbb{R}$, and it is independent of W. Since κ and γ do not play an essential role, for simplicity of the notation, we assume that $\kappa = \frac{1}{2}$ and $\gamma = 1$ in the rest of this chapter.

For $k \in \mathbb{R}$, let space H_k with norm $\|\cdot\|_k$ be defined as in Section 3.3. Throughout this chapter, we will assume that the initial measure X_0 to have a bounded density $v \in H_1$ and we will refer to this condition as *Assumption (I)*.

We recall that

$$\|h\|_{1,\infty}^2 = \sup_x \int_U \left(|h(y,x)|^2 + |\nabla h(y,x)|^2\right) \mu(dy).$$

The following is the main result of this chapter.

Theorem 6.1.1. *Suppose that $h \in \mathbb{H}_2$, $\|h\|_{1,\infty}^2 < 2\kappa$ and X_0 satisfies Assumption (I). Then, the SPRE X_t has a density $v_t(x)$ which is almost surely jointly Hölder continuous. Furthermore, for fixed t its Hölder exponent in x is in $(0, 1/2)$ and for fixed x this exponent in t is in $(0, 1/10)$.*

We now describe the major difficulties to be overcome and sketch the approaches to the main result. When $h = 0$, X becomes the well known Dawson-Watanabe process with the joint continuity for its density studied by Konno and Shiga (1988) and Reimers (1989). As we presented in Chapter 1, this joint continuity is obtained via a convolution technique. If we adopt the same technique here, then the density random field can be represented as

$$v_t(x) = \int_{\mathbb{R}} p_t(x-y)v(y)dy + \int_0^t \int_{\mathbb{R}} \sqrt{v_s(y)}p_{t-s}(x-y)B(dsdy)$$
$$+ \int_0^t \int_U \int_{\mathbb{R}} h(y,z)v_s(z)\partial_z p_{t-s}(x-z)dzW(dsdy), \qquad (6.1.2)$$

where $p_t(x)$ is the heat kernel with generator $\frac{1}{2}\Delta$. However, the third term on the RHS of the above equation is (for some suitable function g) roughly equal to

$$\int_0^t \int_U (t-s)^{-1/2} g(z)W(dsdz),$$

which does *not* converge since

$$\int_0^t \int_U (t-s)^{-1} g(z)\mu(dz)ds = \infty.$$

Therefore, the convolution argument of Konno and Shiga fails in our current model. It actually means that SPDE (6.1.1) does not have a *mild* solution.

Since it is the term containing W that causes the problem, we want to absorb it by the kernel by considering a stochastic transition function. For this purpose let $p^W(s,x;t,y)$ be the random transition function defined by the linear equation (see Lemma 6.3.1 below). We will prove that

$$v_t(y) = \int_{\mathbb{R}} p^W(0,x;t,y)v(x)dx$$
$$+ \int_0^t \int_{\mathbb{R}} \sqrt{v_s(x)}p^W(s,x;t,y)B(dsdx). \qquad (6.1.3)$$

The first term in the above equation is easy to deal with. So we focus on the second term. We will apply Kolmogorov's criteria to obtain the joint continuity. To this end, we need the following estimates to be proved later: for any $y_1,\ y_2 \in \mathbb{R}$ and $t \geq 0$,

$$\mathbb{E}\left|\int_0^t \int_{\mathbb{R}} |p^W(s,x,t,y_1) - p^W(s,x,t,y_2)|^2 dxds\right|^p \leq K|y_1 - y_2|^{2+\epsilon} \quad (6.1.4)$$

and for $y \in \mathbb{R}$ and $t_1 < t_2$,

$$\mathbb{E}\left|\int_0^{t_1}\int_{\mathbb{R}} |p^W(s,x,t_2,y) - p^W(s,x,t_1,y)|^2 dxds\right|^p \leq K|t_1 - t_2|^{2+\epsilon}, \quad (6.1.5)$$

and

$$\mathbb{E}\left|\int_{t_1}^{t_2}\int_{\mathbb{R}} |p^W(s,x,t_1,y)|^2 dxds\right|^p \leq K|t_1 - t_2|^{2+\epsilon},$$

for some $\epsilon > 0$ and suitable $p > 0$.

To obtain (6.1.4) we fix t and let

$$u_s(x) = p^W(t-s,x,t,y_1) - p^W(t-s,x,t,y_2).$$

Then u satisfies the linear SPDE (3.3.1). Estimate (6.1.4) can be obtained using the one we derived in Section 3.3.

For (6.1.5), we note that

$$\tilde{u}_s(x) = p^W(t_1-s,x,t_2,y) - p^W(t_1-s,x,t_1,y)$$

is a solution to the linear SPDE (3.3.1) with initial condition

$$\tilde{u}_0 = p^W(t_1,\cdot,t_2,y) - \delta_y.$$

The LHS of (6.1.5) is then bounded by $\mathbb{E}\|\tilde{u}_0\|_{-1}^{2p}$. To estimate this quantity, we further define

$$u_t(x) = p^W(t_2-t,x,t_2,y)$$

which solves SPDE (3.3.1) with initial $u_0(x) = \delta_y(x)$, and then estimate $\mathbb{E}\|u_{t_2-t_1} - \delta_y\|_{-1}^{2p}$. Similar to what we mentioned above for convolution (6.1.2), we cannot directly apply the convolution with kernel p_t to (3.3.1). We shall use a partial convolution by kernel p_{t^α} where $\alpha \in (0,1)$ is a constant to be decided later. Then

$$\begin{aligned} u_t(z) = {} & p_{t^\alpha}(z-y) + \frac{1}{2}\int_0^t\int_{\mathbb{R}} \Delta u_{t-r}(x) p_{r^\alpha}(z-x)dxdr \\ & + \int_0^t\int_U\int_{\mathbb{R}} \nabla u_{t-r}(x) h(y,x) p_{r^\alpha}(z-x)dxW(dydr) \\ & - \frac{\alpha}{2}\int_0^t\int_{\mathbb{R}} \Delta u_{t-r}(x) p_{r^\alpha}(z-x)dxr^{\alpha-1}dr. \end{aligned} \quad (6.1.6)$$

The main difficulty now lies in the fourth term on the RHS of (6.1.6) because, due to the lack of integrability, we cannot apply integration by parts to move Δ completely to p. Instead, we have to transform a fraction Δ^β of Δ to the kernel p_{r^α} with $\beta < 1$ to be decided (together with α).

The rest of this chapter is organized as follows. In Section 6.2, we prove the existence of the density field and derive the SPDE it satisfies. Then, in Section 6.3 we establish the representation of density $v_t(x)$ in terms of the random transition function. Based on this representation, we estimate the spatial-increments of $v_t(x)$ in Section 6.4 and the time-increments in Section 6.5. We conclude the proof of Theorem 6.1.1 in Section 6.5.

6.2 Derivation of SPDE

In this section, we prove that the SPRE takes values in the space of finite measures which are absolutely continuous with respect to the Lebesgue measure and derive the SPDE satisfied by the density random field.

As in the proof of Theorem 1.4.2, we need the first and second order moments of the process. To this end, we need to introduce the following notation. Let $q_t^{(1)}(x,y)$ be the transition density function of the Markov process $\xi_\alpha(t)$ describing the motion of a typical particle $\alpha \in \mathcal{I}$ in the system. Namely, $\xi_\alpha(t)$ is governed by the following SDE

$$d\xi_\alpha(t) = dB_\alpha(t) + \int_U h(y, \xi_\alpha(t))W(dtdy). \tag{6.2.1}$$

For $\beta \in \mathcal{I}$, let $\xi_\beta(t)$ be the solution of (6.2.1) with driving Brownian motion $B_\alpha(t)$ replaced by an independent copy $B_\beta(t)$. Then, $(\xi_\alpha(t), \xi_\beta(t))$ describes the movement of two typical particles in the system. Let $q_t^{(2)}((x_1,x_2),(y_1,y_2))$ be the transition density function of the 2-dimensional Markov process $(\xi_\alpha(t), \xi_\beta(t))$.

Making use of the moment duality, it is easy to prove that

Lemma 6.2.1. *For any $f,\ g \in C_b(\mathbb{R})$, we have*

$$\mathbb{E}\langle X_t, f\rangle = \int_{\mathbb{R}} \nu(dx) \int_{\mathbb{R}} q_t^{(1)}(x,y) f(y)dy \tag{6.2.2}$$

and

$$\begin{aligned}
&\mathbb{E}\langle X_t, f\rangle \langle X_t, g\rangle \\
&= \int_{\mathbb{R}} \nu(dx_1) \int_{\mathbb{R}} \nu(dx_2) \int_{\mathbb{R}}\int_{\mathbb{R}} q_t^{(2)}((x_1,x_2),(y_1,y_2)) f(y_1) g(y_2) dy_1 dy_2 \\
&\quad + \gamma \int_0^t ds \int_{\mathbb{R}} \nu(dx) \int_{\mathbb{R}} dy q_{t-s}^{(1)}(x,y) \\
&\qquad \times \int_{\mathbb{R}}\int_{\mathbb{R}} q_s^{(2)}((y,y),(z_1,z_2)) f(z_1) g(z_2) dz_1 dz_2.
\end{aligned} \tag{6.2.3}$$

Proof. Equality (6.2.2) follows from the duality right away, so we only prove the second order moment formula (6.2.3). Letting $n_0 = 2$ and

$$f_0(x,y) = f(x)g(y),$$

we define the dual process as that in Section 2.2. Then,

$$\begin{aligned}
\mathbb{E}\left(\langle X_t, f\rangle \langle X_t, g\rangle\right) &= \mathbb{E}\left\langle X_t^{\otimes 2}, f_0\right\rangle \\
&= \mathbb{E}F(\nu, (n_t, f_t)) \exp\left(\frac{\gamma}{2}\int_0^t n_s(n_s-1)ds\right) \\
&= \mathbb{E}1_{\tau_1 > t}\left\langle \nu^{\otimes 2}, f_t\right\rangle e^{\gamma t} + \mathbb{E}1_{\tau_1 \le t}\left\langle \nu, f_t\right\rangle e^{\gamma\tau_1} \\
&= \left\langle \nu^{\otimes 2}, T_t^{(2)} f_0\right\rangle + \gamma\int_0^t \mathbb{E}\left\langle \nu, T_{t-s}^{(1)} S_1 T_s^{(2)} f_0\right\rangle ds,
\end{aligned}$$

where $T_t^{(1)}$ and $T_t^{(2)}$ are the semigroups of the Markov process describing the movements of one and two particles in the system, respectively. Identity (6.2.3) then follows by writing out the semigroups according to the transition density functions. □

The following result is the counterpart of Theorem 1.4.2 in the setting of SPRE.

Theorem 6.2.2. *Suppose that X_0 has density $v_0 \in H_0$. Then, there exists random field $v \in L^2(\Omega \times [0,T] \times \mathbb{R}, dPdxdt)$ such that*

$$\lim_{\epsilon\to 0+} \mathbb{E}\int_0^T \int_{\mathbb{R}} |T_\epsilon^{(1)} X_t(x) - v_t(x)|^2 dxdt = 0. \tag{6.2.4}$$

As a consequence, X_t has density v_t with respect to the Lebesgue measure, and for any $t > 0$,

$$X_t(dx) = v_t(x)dx, \qquad a.s.$$

Proof. Take $f = q_\epsilon^{(1)}(x,\cdot)$ and $g = q_{\epsilon'}^{(1)}(x,\cdot)$ in (6.2.3) and note that as $\epsilon, \epsilon' \to 0$,

$$\begin{aligned}
&\int_{\mathbb{R}^2} q_\epsilon^{(1)}(x,z_1) q_{\epsilon'}^{(1)}(x,z_2) q_{t-s}^{(1)}(z,y) q_t^{(2)}((y,y),(z_1,z_2)) dz_1 dz_2 \\
&\to q_{t-s}^{(1)}(z,y) q_t^{(2)}((y,y),(x,x)).
\end{aligned}$$

By Theorem 6.4.5 in Friedman (1975), there exist constants K and K' such that

$$q_\epsilon^{(1)}(x,y) \le K p_{K'\epsilon}(x-y), \tag{6.2.5}$$

$$q_s^{(2)}((y,y),(z_1,z_2)) \le K p_{K's}(y,z_1) p_{K's}(y,z_2) \tag{6.2.6}$$

where we recall that $p_t(x)$ is the normal density with mean 0 and variance t. Notice that K' is a constant which is usually greater than 1. Since it does not play an essential role, to simplify the notation, we assume $K' = 1$ throughout this section. Hence,

$$\begin{aligned}
&\int_{\mathbb{R}^2} q_\epsilon^{(1)}(x, z_1) q_{\epsilon'}^{(1)}(x, z_2) q_{t-s}^{(1)}(z, y) q_s^{(2)}((y, y), (z_1, z_2)) dz_1 dz_2 \\
&\leq K_1 \int_{\mathbb{R}^2} p_\epsilon(x - z_1) p_{\epsilon'}(x - z_2) p_{t-s}(z - y) p_s(y, z_1) p_s(y, z_2) dz_1 dz_2 \\
&= K_1 p_{s+\epsilon}(x - y) p_{s+\epsilon'}(x - y) p_{t-s}(z - y).
\end{aligned}$$

As

$$\begin{aligned}
&\lim_{\epsilon, \epsilon' \to 0} \int_0^T dt \int dx \int_0^t ds \int_{\mathbb{R}^2} p_{s+\epsilon}(x - y) \\
&\qquad\qquad \times p_{s+\epsilon'}(x - y) p_{t-s}(z - y) dy \nu(dz) \\
&= \lim_{\epsilon, \epsilon' \to 0} \int_0^T dt \int_0^t ds p_{2s+\epsilon+\epsilon'}(0) \nu(\mathbb{R}) \\
&= \int_0^T dt \int_0^t ds p_{2s}(0) \nu(\mathbb{R}) \\
&= \int_0^T dt \int dx \int_0^t ds \int_{\mathbb{R}^2} p_{t-s}(z - y) p_s(x - y) p_s(x - y) dy \nu(dz),
\end{aligned}$$

by the dominated convergence theorem, we see that as $\epsilon, \epsilon' \to 0$,

$$\begin{aligned}
&\int_0^T dt \int dx \int_0^t ds \int_{\mathbb{R}^4} q_{t-s}^{(1)}(z, y) q_\epsilon^{(1)}(x, z_1) \\
&\qquad \times q_{\epsilon'}^{(1)}(x, z_2) q_s^{(2)}((y, y), (z_1, z_2)) dz_1 dz_2 dy \nu(dz) \\
&\to \int_0^T dt \int dx \int_0^t ds \int_{\mathbb{R}^2} q_{t-s}^{(1)}(z, y) q_t^{(2)}((y, y), (x, x)) dy \nu(dz).
\end{aligned}$$

Similarly, as ϵ, $\epsilon' \to 0$, we have

$$\begin{aligned}
&\int_0^T dt \int dx \int_{\mathbb{R}^4} q_\epsilon^{(1)}(x, y_1) q_{\epsilon'}^{(1)}(x, y_2) \\
&\qquad \times q_t^{(2)}((x_1, x_2), (y_1, y_2)) dy_1 dy_2 \nu(dx_1) \nu(dx_2) \\
&\to \int_0^T dt \int dx \int_{\mathbb{R}^2} q_t^{(2)}((x_1, x_2), (x, x)) \nu(dx_1) \nu(dx_2).
\end{aligned}$$

Hence, as ϵ, $\epsilon' \to 0$,

$$\begin{aligned}
&\int_0^T dt \int dx \mathbb{E}\left(\langle X_t, p_\epsilon(x, \cdot)\rangle \langle X_t, p_{\epsilon'}(x, \cdot)\rangle\right) \\
&\to \int_0^T dt \int dx \int_{\mathbb{R}^2} q_t^{(2)}((x_1, x_2), (x, x)) \nu(dx_1) \nu(dx_2) \\
&\quad + \int_0^T dt \int dx \int_0^t ds \int_{\mathbb{R}^2} q_{t-s}^{(1)}(x, y) q_t^{(2)}((y, y), (x, x)) dy \nu(dx).
\end{aligned}$$

From this, we can show that $\{\langle X_t, q^{(1)}(\frac{1}{n}, x, \cdot)\rangle : n = 1, 2, \cdots\}$ is a Cauchy sequence in $L^2(\Omega \times [0,T] \times \mathbb{R})$. This implies the existence of density $v_t(x)$ of X_t in $L^2(\Omega \times [0,T] \times \mathbb{R})$. □

Now we are ready to derive SPDE (6.1.1). We will use the same arguments as those in the proof of Theorem 1.4.3 based on the CMP (2.3.1, 2.3.2) instead of MP (1.1.8, 1.1.9).

Theorem 6.2.3. *The density random field $v_t(x)$ of the SPRE is a solution to the following SPDE:*

$$\partial_t v_t(x) = \frac{1}{2}\Delta v_t(x) - \int_U \nabla(h(y,x)v_t(x))\dot{W}_{ty}dy + \sqrt{v_t(x)}\dot{B}_{tx}. \tag{6.2.7}$$

Namely, $v_t(x)$ is a weak solution of SPDE (6.2.7) in the following sense: for any $f \in C_0^2(\mathbb{R})$, we have

$$\langle v_t, f\rangle_0 = \langle v_0, f\rangle_0 + \int_0^t \left\langle v_s, \frac{1}{2}f''\right\rangle_0 ds + \int_0^t \int_U \langle v_s, h(y,\cdot)f'\rangle_0 W(dsdy)$$
$$+ \int_0^t \int_{\mathbb{R}} \sqrt{v_s(x)} f(x) B(dsdx). \tag{6.2.8}$$

Proof. By the CMP, N_t is a P^W-martingale with quadratic variation process

$$\langle N(f)\rangle_t = \gamma \int_0^t \langle X_s, f^2\rangle ds.$$

Replacing M_t by N_t in the proof of Theorem 1.4.3, we see that there exists a white noise random measure B on $\mathbb{R}_+ \times \mathbb{R}$ independent of W such that

$$N_t(f) = \int_0^t \int_{\mathbb{R}} \sqrt{\gamma v_s(x)} f(x) B(dsdx).$$

Together with CMP (2.3.1, 2.3.2) we see that SPDE (6.2.8) is satisfied. □

6.3 A convolution representation

In this section, we establish a convolution representation for density $v_t(x)$ in terms of a random transition function.

We recall the dual equation on $C_b(\mathbb{R})$:

$$T_{r,t}(x) = f(x) + \int_r^t \kappa\Delta T_{s,t}(x)ds + \int_r^t \int_{\mathbb{R}} \nabla T_{s,t}(x)h(y,x)W(\hat{d}sdy), \tag{6.3.1}$$

where $\hat{d}s$ stands for the backward Itô integral. We denote $T_{r,t}(x)$ by $T^f_{r,t}(x)$ to indicate its dependence on f.

Lemma 6.3.1. *There exists a transition kernel $p^W(s,x;t,\cdot) \in \mathcal{P}(\mathbb{R})$ such that for any $f \in C_b(\mathbb{R})$, we have*

$$T^f_{s,t}(x) = \int_{\mathbb{R}} f(y)p^W(s,x;t,dy). \tag{6.3.2}$$

Furthermore, $p^W(s,x;t,\cdot)$ is absolutely continuous with respect to Lebesgue measure. More specifically, there exists a measurable function $p^W(s,x;t,y)$ such that

$$p^W(s,x;t,A) = \int_A p^W(s,x;t,y)dy, \qquad \forall A \in \mathcal{B}(\mathbb{R}), \ \ a.s.$$

Proof. Taking $\nu = \delta_x$ and replacing the initial time by s, SPDE (3.0.1) has a measure-valued solution. Denote it as

$$p^{s,x,W}_t = P^W(s,x;t,\cdot).$$

The identity (6.3.2) follows from Theorem 3.2.5 directly.

If $f = 1$, it is clear that $T^f_{s,t} = 1$ and hence, $p^{s,x,W}_t \in \mathcal{P}(\mathbb{R})$. Furthermore, applying Theorem 3.3.3, we see that $p^{s,x,W}_t$ has density. □

The following convolution representation is the key in proving the joint continuity of $v_t(y)$. We shall denote $Z(dsdx) \equiv \sqrt{v_s(x)}B(dsdx)$.

Lemma 6.3.2. *Suppose that ν satisfies Assumption (I) and $f \in C^2_b(\mathbb{R})$, then*

$$\langle X_t, f\rangle = \langle \nu, T_{0,t}\rangle + \int_0^t \int_{\mathbb{R}} T_{s,t}(x)Z(dsdx). \tag{6.3.3}$$

Proof. Making use of Lemma 6.2.1, we can prove that for any $t > 0$,

$$\mathbb{E}\|v_t\|_0^2 < \infty.$$

In fact, taking $f = g = \delta_x$, we have

$$\begin{aligned}
\mathbb{E}v_t(x)^2 &= \int_{\mathbb{R}} \nu(dx_1)\int_{\mathbb{R}} \nu(dx_2)q^{(2)}((x_1,x_2),(x,x)) \\
&\quad + \gamma\int_0^t ds\int_{\mathbb{R}} \nu(dz)\int_{\mathbb{R}} dyq^{(1)}_{t-s}(z,y)q^{(2)}_s((y,y),(x,x)) \\
&\le K_1\left(\int_{\mathbb{R}} p_t(x_1-x)\nu(dx_1)\right)^2 \\
&\quad + K_2\int_0^t ds\int_{\mathbb{R}} \nu(dz)\int_{\mathbb{R}} dyp_{t-s}(z-y)p_s(x-y)^2.
\end{aligned}$$

Therefore,

$$\begin{aligned}\mathbb{E}\|v_t\|_0^2 &\le K_1 \int_{\mathbb{R}} dx \int_{\mathbb{R}} p_t(x_1 - x)^2 \nu(dx_1)\nu(\mathbb{R}) \\ &\quad + K_2 \int_0^t ds \int_{\mathbb{R}} \nu(dz) \int_{\mathbb{R}} dy p_{t-s}(z-y) p_{2s}(0) \\ &\le K_1 p_{2t}(0)\nu(\mathbb{R})^2 + K_2 \int_0^t p_{2s}(0) ds \nu(\mathbb{R}) < \infty.\end{aligned}$$

Denote the RHS of (6.3.3) by $\langle Y_t, f\rangle$. It is easy to show that Y_t is an H_0-valued process. For $f \in C_b^2(\mathbb{R})$ we define

$$\begin{aligned}M_t \equiv \langle Y_t, f\rangle - \langle X_0, f\rangle - \int_0^t \left\langle Y_s, \frac{1}{2}\Delta f\right\rangle ds \\ - \int_0^t \int_U \langle Y_s, h(y,\cdot)\nabla f\rangle W(dsdy).\end{aligned}$$

By (6.3.3), we see that

$$\begin{aligned}M_t &= \left\langle X_0, T_{0,t}^f\right\rangle + \int_0^t \int_{\mathbb{R}} T_{s,t}^f(x) Z(dsdx) - \langle X_0, f\rangle \\ &\quad - \int_0^t \left\{ \left\langle X_0, T_{0,s}^{\frac{1}{2}\Delta f}\right\rangle + \int_0^s \int_{\mathbb{R}} T_{r,s}^{\frac{1}{2}\Delta f}(x) Z(drdx)\right\} ds \\ &\quad - \int_0^t \int_U \left\{ \left\langle X_0, T_{0,s}^{h(y,\cdot)\nabla f}\right\rangle + \int_0^s \int_{\mathbb{R}} T_{r,s}^{h(y,\cdot)\nabla f}(x) Z(drdx)\right\} W(dsdy) \\ &\equiv \langle X_0, f_t\rangle + \int_0^t \int_{\mathbb{R}} g_{s,t}(x) Z(dsdx),\end{aligned}$$

where

$$f_t = T_{0,t}^f - f - \int_0^t T_{0,s}^{\frac{1}{2}\Delta f} ds - \int_0^t \int_U T_{0,s}^{h(y,\cdot)\nabla f} W(dsdy)$$

and

$$g_{s,t}(x) = T_{s,t}^f(x) - \int_s^t T_{s,r}^{\frac{1}{2}\Delta f}(x) dr - \int_s^t \int_U T_{s,r}^{h(y,\cdot)\nabla f}(x) W(drdy).$$

By the definition of $T_{s,t}^f$ and Theorem 3.2.5, we see that $f_t = 0$ and $g_s(x) = f(x)$. Hence,

$$M_t = \int_0^t \int_{\mathbb{R}} f(x) Z(dsdx).$$

Let $\tilde{X}_t = X_t - Y_t$. By (6.1.1), $\tilde{X}$ is an H_0-valued solution to the following linear SDE

$$\left\langle \tilde{X}_t, f\right\rangle = \int_0^t \left\langle \tilde{X}_s, \frac{1}{2}\Delta f\right\rangle ds + \int_0^t \int_{\mathbb{R}} \left\langle \tilde{X}_s, h(y,\cdot)\nabla f\right\rangle W(dsdy). \quad (6.3.4)$$

Using Theorem 3.1.6 we can conclude that $\tilde{X} = 0$. □

Remark 6.3.3. The identity (6.1.3) follows by plugging (6.3.2) into equation (6.3.3). The random field satisfying (6.1.3) is called a conditional mild solution of SPDE (6.2.7).

6.4 An estimate in spatial increment

In this section we estimate the spatial increment of density $v_t(y)$. As a consequence, we will see that for $t > 0$ fixed, $v_t(y)$ is Hölder continuous in y with exponent $1/2 - \epsilon$.

By Lemma 6.3.2, $v_t(y)$ can be represented as

$$\begin{aligned} v_t(y) &= \int_{\mathbb{R}} \nu(x)p^W(0,x;t,y)dx + \int_0^t \int_{\mathbb{R}} p^W(s,x;t,y)Z(dsdx) \\ &\equiv v_t^1(y) + v_t^2(y). \end{aligned} \tag{6.4.1}$$

To prove joint continuity by Kolmogorov's criteria, we need the estimate given in Lemma 6.4.2 below. First, we present the finiteness statement about the moment of the density random field. The proof follows from the same argument as that in the proof of Lemma 1.4.5 using the fact (6.2.5) and the moment formula in Theorem 2.2.2.

Lemma 6.4.1. *If ν is finite and satisfies*

$$\sup_{t,x} \langle \nu, p_t(x-\cdot)\rangle < \infty, \tag{6.4.2}$$

then

$$\sup_{t>0,\ x\in\mathbb{R}} \mathbb{E}v_t(x)^n < \infty \tag{6.4.3}$$

for all $n \in \mathbb{N}$.

Lemma 6.4.2. *Suppose that Assumption (I) holds. Then $\forall\ p \geq 1$, there exists a constant K_1 such that*

$$\begin{aligned} &\mathbb{E}\left|\int_0^t \int_{\mathbb{R}} (p^W(s,x;t,y_1) - p^W(s,x;t,y_2))Z(dsdx)\right|^{2p} \\ &\leq K_1 \left(\mathbb{E}\left|\int_0^t \int_{\mathbb{R}} (p^W(s,x;t,y_1) - p^W(s,x;t,y_2))^2 dxds\right|^{2p-1}\right)^{\frac{p}{2p-1}}. \end{aligned} \tag{6.4.4}$$

Proof. By Burkholder inequality, we have

$$\begin{aligned} L &\equiv \mathbb{E}\left|\int_0^t \int_{\mathbb{R}} (p^W(s,x;t,y_1) - p^W(s,x;t,y_2))Z(dsdx)\right|^{2p} \\ &\leq K_2 \mathbb{E}\left|\int_0^t \int_{\mathbb{R}} (p^W(s,x;t,y_1) - p^W(s,x;t,y_2))^2 v_s(x)dxds\right|^{p}. \end{aligned}$$

For $2 = (2p-1)/p + 1/p$, applying Cauchy-Schwartz inequality we obtain

$$\begin{aligned}
L \le& K_2 \mathbb{E}\left(\left| \int_0^t \int_{\mathbb{R}} (p^W(s,x;t,y_1) - p^W(s,x;t,y_2))^2 dx ds \right|^{\frac{2p-1}{2}} \right.\\
&\left. \times \left| \int_0^t \int_{\mathbb{R}} (p^W(s,x;t,y_1) - p^W(s,x;t,y_2))^2 v_s(x)^{2p} dx ds \right|^{\frac{1}{2}} \right)\\
\le& K_2 \left(\mathbb{E} \left| \int_0^t \int_{\mathbb{R}} (p^W(s,x;t,y_1) - p^W(s,x;t,y_2))^2 dx ds \right|^{2p-1} \right)^{\frac{1}{2}}\\
&\times \left(\mathbb{E} \int_0^t \int_{\mathbb{R}} (p^W(s,x;t,y_1) - p^W(s,x;t,y_2))^2 v_s(x)^{2p} dx ds \right)^{\frac{1}{2}}\\
\equiv& K_2 I \times J.
\end{aligned}$$

It then follows from Lemma 6.4.1 that

$$\begin{aligned}
J \le& K_3 \left(\mathbb{E} \int_0^t \int_{\mathbb{R}} (p^W(s,x;t,y_1) - p^W(s,x;t,y_2))^2 dx ds \right)^{\frac{1}{2}}\\
=& K_3 I^{1/(2p-1)}.
\end{aligned}$$

Thus, $L \le K_1 I^{2p/(2p-1)}$ which coincides with the RHS of (6.4.4). □

As a consequence of Theorem 3.3.3, we get the following result.

Proposition 6.4.3. *Suppose that the conditions of Theorem 6.1.1 hold. Let $t \in [0,T]$ and $p \ge 1$ be fixed. Then, there exists a constant $K = K(p,T)$ such that*

$$\mathbb{E}|v_t^2(y_1) - v_t^2(y_2)|^{2p} \le K|y_1 - y_2|^p, \qquad \forall\, y_1,\ y_2 \in \mathbb{R}. \tag{6.4.5}$$

Consequently, for $t > 0$ fixed, v_t^2 is Hölder continuous with exponent $1/2 - \epsilon$ for any $\epsilon > 0$.

Proof. Let

$$u_s(x) = p_{t-s}^W(x,t,y_1) - p_{t-s}^W(x,t,y_2).$$

Then u solves equation (3.3.1) with $u_0 = \delta_{y_1} - \delta_{y_2}$. For any $f \in H_1$ we have

$$|\langle u_0, f\rangle| = |f(y_1) - f(y_2)| = \left| \int_{y_1}^{y_2} f'(s) ds \right| \le \sqrt{|y_2 - y_1|}\|f\|_1.$$

Thus, $u_0 \in H_{-1}$ and

$$\|u_0\|_{-1} \le \sqrt{|y_2 - y_1|}. \tag{6.4.6}$$

By Theorem 3.3.3 we get

$$\mathbb{E}\left(\int_0^t \int_{\mathbb{R}} |p^W(s,x,t,y_1) - p^W(s,x,t,y_2)|^2 dxds\right)^p \leq K_1|y_1 - y_2|^p.$$

Inequality (6.4.5) then follows from Lemma 6.4.2. □

Finally, we consider $v_t^1(y)$.

Proposition 6.4.4. *Suppose that the conditions of Theorem 6.1.1 hold. Then, for $p \geq 1$ and $T > 0$, there exists a constant $K = K(p,T)$ such that*

$$\mathbb{E}|v_t^1(y_1) - v_t^1(y_2)|^{2p} \leq K|y_1 - y_2|^p, \qquad \forall\, t \in [0,T].$$

Proof. Note that

$$\begin{aligned}\mathbb{E}|v_t^1(y_1) - v_t^1(y_2)|^{2p} &= \mathbb{E}\left|\int_{\mathbb{R}} \left(p^W(0,x;t,y_1) - p^W(0,x;t,y_2)\right)\nu(x)dx\right|^{2p} \\ &\leq \mathbb{E}\|p^W(0,\cdot;t,y_1) - p^W(0,\cdot;t,y_2)\|_{-1}^{2p}\|\nu\|_1^{2p} \\ &\leq K_1\|\delta_{y_1} - \delta_{y_2}\|_{-1}^{2p}\|\nu\|_1^{2p},\end{aligned}$$

where the first inequality is a consequence of the duality between Hilbert spaces H_{-1} and H_1; and the second inequality is implied by Theorem 3.3.3. The conclusion then follows from (6.4.6). □

6.5 Estimates in time increment

In this section we consider time-increments of the types of

$$\int_0^{t_1} \int_{\mathbb{R}} \left(p^W(s,x;t_2,y) - p^W(s,x;t_1,y)\right) Z(dsdx) \tag{6.5.1}$$

and

$$\int_{t_1}^{t_2} \int_{\mathbb{R}} p^W(s,x;t_2,y) Z(dsdx). \tag{6.5.2}$$

For the type of (6.5.1), we first use Theorem 3.3.3 to obtain a preliminary estimate by $\mathbb{E}\|u_{t_2-t_1} - \delta_y\|_{-1}^{2p}$, where u_t is a solution to SDE (3.3.1) with $u_0 = \delta_y$. To further estimate this quantity, we need to develop two major techniques, i.e., the partial convolution by kernel p_{r^α} and the partial integration by parts introduced in Section 6.1. For the type of (6.5.2), we will use a technique developed by Xiong and Zhou (2004).

Lemma 6.5.1. *For any $t_1 < t_2$ and $y \in \mathbb{R}$, we have*

$$\mathbb{E}\left(\int_0^{t_1} \int_{\mathbb{R}} \left(p^W(s,x;t_2,y) - p^W(s,x;t_1,y)\right) Z(dsdx)\right)^{2p}$$
$$\leq K\mathbb{E}\|p^W(t_1,\cdot;t_2,y) - \delta_y\|_{-1}^{2p}.$$

Proof. Note that

$$\tilde{u}_s(x) \equiv p^W(t_1 - s, x; t_2, y) - p^W(t_1 - s, x; t_1, y)$$

is the solution of SPDE (3.3.1) with initial condition $p^W(t_1, \cdot; t_2, y) - \delta_y$ and hence,

$$\begin{aligned}
&\mathbb{E}\left(\int_0^{t_1}\int_{\mathbb{R}}\left(p^W(s,x;t_2,y) - p^W(s,x;t_1,y)\right) Z(dsdx)\right)^{2p} \\
&\leq K_1\left(\mathbb{E}\left(\int_0^{t_1}\int_{\mathbb{R}}\left(p^W(s,x;t_2,y) - p^W(s,x;t_1,y)\right)^2 dsdx\right)^{2p-1}\right)^{\frac{p}{2p-1}} \\
&\leq K_2\mathbb{E}\|p^W(t_1,\cdot;t_2,y) - \delta_y\|_{-1}^{2p}.
\end{aligned}$$

□

Let

$$u_s(x) = p^W(t_2 - s, x; t_2, y).$$

Then u solves SPDE (3.3.1) with $u_0 = \delta_y$. Since Δu_s is not in H_{-1} we cannot use (3.3.1) directly to get an estimate on

$$\mathbb{E}\|u_{t_2-t_1} - \delta_y\|_{-1}^{2p}.$$

Instead, fixing t and taking differential of

$$\int_{\mathbb{R}} u_{t-r}(x)p_{r^\alpha}(z-x)dx$$

with respect to r, and then taking the integral we get (6.1.6). Denote the second and the third terms on the RHS of (6.1.6) by $\frac{1}{2}I_2$ and I_3, respectively. Write the fourth term by $\frac{\alpha}{2}(I_4 - I_5)$ with

$$I_4 = \int_0^t\int_{\mathbb{R}}(I-\Delta)u_{t-r}(x)p_{r^\alpha}(z-x)dxr^{\alpha-1}dr$$

and

$$I_5 = \int_0^t\int_{\mathbb{R}} u_{t-r}(x)p_{r^\alpha}(z-x)dxr^{\alpha-1}dr.$$

Then

$$u_t(z) - \delta_y(z) = I_1 + \frac{1}{2}I_2 + I_3 + \frac{\alpha}{2}(I_4 - I_5).$$

We now estimate I_j, $j = 1, 2, \cdots, 5$, separately. Although the following result can be implied directly from the analyticity of Δ on $L^2(\mathbb{R})$, we give a brief and elementary proof for the convenience of the reader.

Lemma 6.5.2. *For $\beta \in (0,1)$ there is a constant such that for $r \in (0,T)$ we have*

$$\int_{\mathbb{R}}\left|(I-\Delta)^\beta p_r(x)\right| dx \leq Kr^{-\beta}. \tag{6.5.3}$$

Proof. Note that the integral in (3.3.2) for the definition of $(I-\Delta)^{\beta}$ can be split into two parts: $J_1(x)$ denotes the part from 0 to r and $J_2(x)$ is from r to ∞. Namely,

$$J_1(x)=c(\beta)\int_0^r\frac{e^{-t}p_{t+r}(x)-p_r(x)}{t^{1+\beta}}dt$$

and

$$J_2(x)=c(\beta)\int_r^{\infty}\frac{e^{-t}p_{t+r}(x)-p_r(x)}{t^{1+\beta}}dt.$$

Then

$$\int_{\mathbb{R}}|J_2(x)|dx\le K\int_r^{\infty}\frac{e^{-t}+1}{t^{1+\beta}}dt\le Kr^{-\beta}.$$

For $t\le r$, we have

$$\begin{aligned}&\left|e^{-t}p_{t+r}(x)-p_r(x)\right|p_{t+r}(x)^{-1}\\&=\left|e^{-t}-\sqrt{\frac{t+r}{r}}\exp\left(-\frac{x^2}{2r}+\frac{x^2}{2(t+r)}\right)\right|\\&\le\left|e^{-t}-1\right|+\left|1-\sqrt{\frac{t+r}{r}}\right|+\sqrt{\frac{t+r}{r}}\left|1-\exp\left(-\frac{tx^2}{2r(t+r)}\right)\right|\\&\le\sqrt{2}\left(\frac{tx^2}{2r(r+t)}+t+\frac{t}{r}\right).\end{aligned}$$

Multiplying both sides by $p_{t+r}(x)$ and taking the integral we see that

$$\int_{\mathbb{R}}|J_1(x)|dx\le Kr^{-\beta}.$$

□

Now we estimate I_4. Note that

$$\begin{aligned}&\|I_4\|_{-1}\\&\le\int_0^t\left\|\int_{\mathbb{R}}(I-\Delta)^{\frac{1+\beta}{2}}u_{t-r}(\cdot-x)(I-\Delta)^{\frac{1-\beta}{2}}p_{r^{\alpha}}(x)dx\right\|_{-1}r^{\alpha-1}dr\\&\le K\int_0^t\left\|(I-\Delta)^{\frac{1+\beta}{2}}u_{t-r}\right\|_{-1}\int_{\mathbb{R}}\left|(I-\Delta)^{\frac{1-\beta}{2}}p_{r^{\alpha}}(x)\right|dxr^{\alpha-1}dr\\&\le K\int_0^t\left\|(I-\Delta)^{\frac{1+\beta}{2}}u_{t-r}\right\|_{-1}r^{-\frac{\alpha}{2}(1-\beta)}r^{\alpha-1}dr\\&\le K\left(\int_0^t\left\|(I-\Delta)^{\frac{1+\beta}{2}}u_{t-r}\right\|_{-1}^2dr\right)^{\frac{1}{2}}\left(\int_0^t r^{\alpha(1+\beta)-2}dr\right)^{\frac{1}{2}}\\&=K\left(\int_0^t\|u_r\|_{\beta}^2dr\right)^{\frac{1}{2}}t^{(\alpha(1+\beta)-1)/2}\end{aligned}$$

where $\beta \in (0, 1/2)$ is chosen such that $\alpha(1+\beta) > 1$. Thus,

$$\mathbb{E}\|I_4\|_{-1}^{2p} \leq Kt^{(\alpha(1+\beta)-1)p}.$$

Terms I_2 and I_5 can be estimated similarly (less difficult).

Next, we estimate I_3. Note that

$$\begin{aligned}
&\int_0^t \int_U \left\| \int_{\mathbb{R}} \nabla u_{t-r}(x) h(y,x) p_{r^\alpha}(\cdot - x) dx \right\|_{-1}^2 dy dr \\
=& \int_0^t \int_U \left\| \int_{\mathbb{R}} u_{t-r}(x) \nabla h(y,x) p_{r^\alpha}(\cdot - x) dx \right\|_{-1}^2 dy dr \\
&+ \int_0^t \int_U \left\| \int_{\mathbb{R}} u_{t-r}(x) h(y,x) \nabla p_{r^\alpha}(\cdot - x) dx \right\|_{-1}^2 dy dr \\
\equiv& I_{31} + I_{32}.
\end{aligned}$$

Term I_{32} can be calculated by the following

$$\begin{aligned}
I_{32} =& \int_0^t \int_U \left\| \int_{\mathbb{R}} u_{t-r}(\cdot - x) h(y, x - \cdot) \nabla p_{r^\alpha}(x) dx \right\|_{-1}^2 dy dr \\
=& \int_0^t \int_U \int_{\mathbb{R}} \int_{\mathbb{R}} \langle u_{t-r}(\cdot - x) h(y, x - \cdot), u_{t-r}(\cdot - x') h(y, x' - \cdot) \rangle_{-1} \\
&\times \nabla p_{r^\alpha}(x) \nabla p_{r^\alpha}(x') dx dx' dy dr \\
=& \int_0^t \int_U \int_{\mathbb{R}} \int_{\mathbb{R}} \int_0^\infty \int_0^\infty (uv)^{-1/2} e^{-(u+v)} \\
&\times \int_{\mathbb{R}} \int_{\mathbb{R}} p_u(z - z_1) u_{t-r}(z_1 - x) h(y, x - z_1) dz_1 \\
&\quad \times \int_{\mathbb{R}} p_v(z - z_2) u_{t-r}(z_2 - x') h(y, x' - z_2) dz_2 dz du dv \\
&\quad\quad \times \nabla p_{r^\alpha}(x) \nabla p_{r^\alpha}(x') dx dx' dy dr.
\end{aligned}$$

Using the definition of $\rho(x, y)$ and its boundedness, we can then estimate I_{32} as follows

$$\begin{aligned}
I_{32} \leq& K \int_0^t \int_{\mathbb{R}} \int_{\mathbb{R}} \langle u_{t-r}(\cdot - x), u_{t-r}(\cdot - x') \rangle_{-1} |\nabla p_{r^\alpha}(x)| |\nabla p_{r^\alpha}(x')| dx dx' dr \\
\leq& K \int_0^t \left(\int_{\mathbb{R}} \|u_{t-r}(\cdot - x)\|_{-1} |\nabla p_{r^\alpha}(x)| dx \right)^2 dr \\
\leq& K \sup_{r \leq t} \|u_r\|_{-1}^2 \int_0^t r^{-\alpha} dr \\
\leq& K \sup_{r \leq t} \|u_r\|_{-1}^2 t^{1-\alpha}.
\end{aligned}$$

I_{31} can be estimated similarly, and the estimation for I_1 is easy to achieve. To summarize, we have the proposition below.

Proposition 6.5.3. *For $p \geq 1$, $\alpha \in (0,1)$ and $\beta \in (0,1/2)$ satisfying $\alpha(1+\beta) > 1$, there exists a constant K such that $\forall\ t_1 < t_2$, we have*

$$\begin{aligned}&\mathbb{E}\left(\int_0^{t_1}\int_{\mathbb{R}}\left(p^W(s,x;t_2,y)-p^W(s,x;t_1,y)\right)^2 Z(dsdx)\right)^p\\&\leq K\max\left(|t_2-t_1|^{(\alpha(1+\beta)-1)p}, |t_2-t_1|^{(1-\alpha)p}\right).\end{aligned}$$

Finally, we estimate

$$\mathbb{E}\left(\int_{t_1}^{t_2}\int_{\mathbb{R}} p^W(s,x,t_2,y)^2 Z(dsdx)\right)^{2p}.$$

Similar to Section 6.4, the above moment is bounded by

$$\left(\mathbb{E}\left(\int_{t_1}^{t_2}\int_{\mathbb{R}} p^W(s,x,t_2,y)^2 dxds\right)^{2p-1}\right)^{\frac{p}{2p-1}}$$

which we shall estimate using the method of Xiong and Zhou (2004).

The key identity is the following lemma.

Lemma 6.5.4. *For any $k \in \mathbb{N}$, $s < t$ and $x, y \in \mathbb{R}^k$, we have*

$$\mathbb{E}\prod_{i=1}^k p^W(s,x_i,t,y_i) = q^{(k)}_{t-s}(x,y),$$

where $q^{(k)}$ is the transition function of the k-dimensional Markov process consisting of the motion of k particles of the branching particle system introduced in Section 2.1.

Proof. Let t and y be fixed. We define

$$u^i_r(x^i) = p^W(t-r,x^i,t,y), \qquad i = 1, 2, \cdots, k.$$

Then u^i is a solution to (3.3.1) with initial δ_y. Applying Itô's formula to the product and taking expectation, we get

$$\frac{d}{dr}\mathbb{E}\prod_{i=1}^k u^i_r(x^i) = \mathcal{A}^{(k)}\mathbb{E}\prod_{i=1}^k u^i_r(x^i)$$

where $\mathcal{A}^{(k)}$ is the generator of the k-dimensional Markov process consisting of the motion of k particles of the branching particles system. The conclusion of the lemma then follows easily. □

Lemma 6.5.5. *For any integer $n \geq 1$ and $y \in \mathbb{R}$, we have*

$$\mathbb{E}\left(\int_{t_1}^{t_2}\int_{\mathbb{R}} p^W(s,x,t_2,y)^2 dxds\right)^n \leq K|t_2 - t_1|^{n/2}. \tag{6.5.4}$$

Proof. Let $t_1 = 0$ and $t_2 = t$ for simplicity and denote the LHS of (6.5.4) by L. Then, we can calculate L as follows.

$$\begin{aligned}
&n!\mathbb{E}\int_0^t ds_1 \int_{s_1}^t ds_2 \cdots \int_{s_{n-1}}^t ds_n \int_{\mathbb{R}}\cdots\int_{\mathbb{R}} dx_1\cdots dx_n \prod_{i=1}^n p^W(s_i,x_i,t,y)^2\\
&= n!\mathbb{E}\int_0^t ds_1 \int_{s_1}^t ds_2 \cdots \int_{s_{n-1}}^t ds_n \int_{\mathbb{R}}\cdots\int_{\mathbb{R}} dx_1\cdots dx_n \prod_{i=2}^n p^W(s_i,x_i,t,y)^2\\
&\quad\times \int_{\mathbb{R}} p^W(s_1,x_1,s_2,x_{11})p^W(s_2,x_{11},t,y)dx_{11}\\
&\qquad\times \int_{\mathbb{R}} p^W(s_1,x_1,s_2,x_{12})p^W(s_2,x_{12},t,y)dx_{12}\\
&= n!\mathbb{E}\int_0^t ds_1 \int_{s_1}^t ds_2 \cdots \int_{s_{n-1}}^t ds_n \int_{\mathbb{R}}\cdots\int_{\mathbb{R}} dx_1\cdots dx_n \prod_{i=2}^n p^W(s_i,x_i,t,y)^2\\
&\quad\times \int_{\mathbb{R}}\int_{\mathbb{R}} q^{(2)}_{s_2-s_1}((x_1,x_1),(x_{11},x_{12}))\\
&\qquad\times p^W(s_2,x_{11},t,y)p^W(s_2,x_{12},t,y)dx_{11}dx_{12},
\end{aligned} \tag{6.5.5}$$

where the last equality follows from Lemma 6.5.4 and the independency of $p^W(s_1,x_1,s_2,x_{11})p^W(s_1,x_1,s_2,x_{12})$ and the other factors, which are measurable with respect to $\mathcal{F}^W_{s_1,s_2}$ and $\mathcal{F}^W_{s_2,t}$, respectively.

Note that

$$\begin{aligned}
q^{(2)}_{s_2-s_1}((x_1,x_1),(x_{11},x_{12})) &\leq K_1 p_{s_2-s_1}(x_1-x_{11})p_{s_2-s_1}(x_1-x_{12})\\
&\leq \frac{K_2}{\sqrt{s_2-s_1}} p_{s_2-s_1}(x_1-x_{11}).
\end{aligned}$$

Using (6.5.5), we now estimate L with

$$\begin{aligned}
L \leq K_3\mathbb{E}\int_0^t \frac{ds_1}{\sqrt{s_2-s_1}}\int_{s_1}^t ds_2\cdots\int_{s_{n-1}}^t ds_n \int_{\mathbb{R}}\cdots\int_{\mathbb{R}} dx_{11}dx_{12}dx_2\cdots dx_n\\
\times p^W(s_2,x_{11},t,y)p^W(s_2,x_{12},t,y)\prod_{i=2}^n p^W(s_i,x_i,t,y)^2.
\end{aligned}$$

We further calculate to obtain

$$
\begin{aligned}
L \leq K_4 \mathbb{E} \int_0^t \frac{ds_1}{\sqrt{s_2 - s_1}} \int_{s_1}^t ds_2 \cdots \int_{s_{n-1}}^t ds_n \int_{\mathbb{R}} \cdots \int_{\mathbb{R}} dx_{11} dx_{12} dx_2 \cdots dx_n \\
\times \int_{\mathbb{R}} p^W(s_2, x_{11}, s_3, x'_{11}) p^W(s_3, x'_{11}, t, y) dx'_{11} \\
\times \int_{\mathbb{R}} p^W(s_2, x_{12}, s_3, x'_{12}) p^W(s_3, x'_{12}, t, y) dx'_{12} \\
\times \int_{\mathbb{R}} p^W(s_2, x_2, s_3, x_{21}) p^W(s_3, x_{21}, t, y) dx_{21} \\
\times \int_{\mathbb{R}} p^W(s_2, x_2, s_3, x_{22}) p^W(s_3, x_{22}, t, y) dx_{22} \\
\times \prod_{i=3}^n p^W(s_i, x_i, t, y)^2.
\end{aligned}
$$

Making use of Lemma 6.5.4, we then have

$$
\begin{aligned}
L \leq K_5 \mathbb{E} \int_0^t \frac{ds_1}{\sqrt{s_2 - s_1}} \int_{s_1}^t ds_2 \cdots \int_{s_{n-1}}^t ds_n \int_{\mathbb{R}} \cdots \int_{\mathbb{R}} dx_{11} dx_{12} dx_2 \cdots dx_n \\
\times \int_{\mathbb{R}} \int_{\mathbb{R}} \int_{\mathbb{R}} \int_{\mathbb{R}} dx'_{11} dx'_{12} dx_{21} dx_{22} \\
q^{(4)}_{s_3 - s_2}((x_{11}, x_{12}, x_2, x_2), (x'_{11}, x'_{12}, x_{21}, x_{22})) \\
\times p^W(s_3, x'_{11}, t, y) p^W(s_3, x'_{12}, t, y) p^W(s_3, x_{21}, t, y) \\
\times p^W(s_3, x_{22}, t, y) \prod_{i=3}^n p^W(s_i, x_i, t, y)^2.
\end{aligned}
$$

Note that

$$
\begin{aligned}
& q^{(4)}_{s_3 - s_2}((x_{11}, x_{12}, x_2, x_2), (x'_{11}, x'_{12}, x_{21}, x_{22})) \\
\leq & \frac{K_6}{\sqrt{s_3 - s_2}} p_{s_3 - s_2}(x'_{11} - x_{11}) p_{s_3 - s_2}(x'_{12} - x_{12}) p_{s_3 - s_2}(x_{21} - x_2).
\end{aligned}
$$

We continue to estimate L with

$$
\begin{aligned}
L \leq K_7 \mathbb{E} \int_0^t \frac{ds_1}{\sqrt{s_2 - s_1}} \int_{s_1}^t \frac{ds_2}{\sqrt{s_3 - s_2}} \cdots \int_{s_{n-1}}^t ds_n \\
\times \int_{\mathbb{R}} \cdots \int_{\mathbb{R}} dx'_{11} dx'_{12} dx_{21} dx_{22} dx_3 \cdots dx_n \\
\times p^W(s_3, x'_{11}, t, y) p^W(s_3, x'_{12}, t, y) p^W(s_3, x_{21}, t, y) \\
\times p^W(s_3, x_{22}, t, y) \prod_{i=3}^n p^W(s_i, x_i, t, y)^2.
\end{aligned}
$$

By continuing this procedure, we see that

$$\begin{aligned}
L &\le K_8 \mathbb{E} \int_0^t \frac{ds_1}{\sqrt{s_2 - s_1}} \int_{s_1}^t \frac{ds_2}{\sqrt{s_3 - s_2}} \cdots \int_{s_{n-1}}^t \frac{ds_n}{\sqrt{t - s_n}} \\
&\quad \times \int_{\mathbb{R}} \cdots \int_{\mathbb{R}} dx_{11} dx_{12} \cdots dx_{n1} dx_{n2} \\
&\quad \times \prod_{i=1}^n p^W(s_n, x_{i1}, t, y) p^W(s_n, x_{i2}, t, y) \\
&\le K_9 \mathbb{E} \int_0^t \frac{ds_1}{\sqrt{s_2 - s_1}} \int_{s_1}^t \frac{ds_2}{\sqrt{s_3 - s_2}} \cdots \int_{s_{n-1}}^t \frac{ds_n}{\sqrt{t - s_n}} \\
&\le K t^{n/2}.
\end{aligned}$$

Thus we finish the proof by replacing t with $t_2 - t_1$. □

To summarize, we get

Proposition 6.5.6. *Suppose that the conditions of Theorem 6.1.1 hold. Then, there exist integer $p \ge 1$ and real numbers $\epsilon > 0$ and $K > 0$ such that $\forall\ t_1 < t_2$ and $y \in \mathbb{R}$, we have*

$$\mathbb{E}|v_{t_1}^2(y) - v_{t_2}^2(y)|^{2p} \le K|t_1 - t_2|^{2+\epsilon}. \tag{6.5.6}$$

Proof. Choose $p \ge 2$, $\alpha \in (0,1)$ and $\beta \in \left(0, \frac{1}{2}\right)$ such that

$$\min\left\{(\alpha(1+\beta) - 1)p,\ (1-\alpha)p,\ \frac{p}{2}\right\} \ge 2 + \epsilon.$$

By Proposition 6.5.3 and Lemma 6.5.5, we see that (6.5.6) holds. □

Note that

$$\begin{aligned}
\mathbb{E}|v_{t_1}^1(y) - v_{t_2}^1(y)|^{2p} &= \mathbb{E}\left|\int_{\mathbb{R}} \left(p^W(0, x; t_2, y) - p^W(0, x; t_1, y)\right) \nu(x) dx\right|^{2p} \\
&\le \mathbb{E}\|p^W(0, \cdot; t_2, y) - p^W(0, \cdot; t_1, y)\|_{-1}^{2p} \|\nu\|_1^{2p}.
\end{aligned}$$

Similar to the proof for $v_t^2(y)$, we get the following

Proposition 6.5.7. *Suppose that the conditions of Theorem 6.1.1 hold. Then, there exist integer $p \ge 1$ and real numbers $\epsilon > 0$ and $K > 0$ such that $\forall\ t_1 < t_2$ and $y \in \mathbb{R}$, we have*

$$\mathbb{E}|v_{t_1}^1(y) - v_{t_2}^1(y)|^{2p} \le K|t_1 - t_2|^{2+\epsilon}.$$

Finally, we are ready to finish the proof of Theorem 6.1.1.

Proof. Combining Propositions 6.4.3, 6.4.4, 6.5.6 and 6.5.7, we get

$$\mathbb{E}|X_{t_1}(y_1) - X_{t_2}(y_2)|^{2p} \leq K|(t_1, y_1) - (t_2, y_2)|^{2+\epsilon}.$$

The joint continuity then follows from Kolmogorov's criteria.

About the Hölder continuity, for fixed t, it follows from Propositions 6.4.3 and 6.4.4 that $u_t(x)$ is Hölder continuous in x with exponent $\frac{1}{2} - \epsilon$.

On the other hand, it follows from Proposition 6.5.3 that $v_t(x)$ is Hölder continuous in t with exponent $\min(\alpha(1+\beta) - 1, 1 - \alpha)/2 - \epsilon$. Since $\alpha < 1$ and $\beta < 1/2$, the best Hölder exponent we can get here is $1/10 - \epsilon$. $\square$

6.6 Historical remarks

This chapter is based on the paper of Li, Wang, Xiong and Zhou (2012) where the case of $U = \mathbb{R}$ and $h(y,x) = \tilde{h}(y - x)$ is studied. In that case, SPDE (6.1.1) is derived by Dawson, Vaillancourt and Wang (2000). The existence of density random field is proved by Wang (1997). The key identity given in Lemma 6.5.4 is taken from Xiong and Zhou (2004).

When the third term on the RHS of (6.1.1) is replaced by $\int_0^t \nabla(h(x)v_s(x))d\tilde{W}(s)$ with a real-valued Brownian motion $\tilde{W}$, the SPDE is satisfied by the density process of a measure-valued process for the model studied by Skoulakis and Adler (2001). For that model, Lee, Mueller and Xiong (2009) proved the continuity in x for almost all fixed t using Krylov's (cf. Krylov (1999)) L_p theory for linear SPDEs.

It is conjectured by Hu and Nualart that for x fixed, $v_t(x)$ should be Hölder continuous in t with exponent $1/4 - \epsilon$. This conjecture is confirmed by Hu, Lu and Nualart (2012) using Malliavin's calculus. We do not include this nice result in this book because we do not assume the familiarity of the Malliavin's calculus by the reader.

Chapter 7

Backward Doubly Stochastic Differential Equations

As we saw in Chapters 1 and 6, the coefficients of the SPDEs arising from superprocesses are usually non-Lipschitz, which make the uniqueness problem for such equations very challenging. In this chapter, we consider a special class of SDEs, called the backward doubly stochastic differential equations (BDSDEs), whose uniqueness can be obtained by Yamada-Watanabe's argument. Furthermore, as we shall see in the next chapter, the uniqueness problem for such BDSDEs is equivalent to that of the corresponding SPDEs.

7.1 Introduction and basic definitions

Let $(\mathbb{U}, \mathcal{B}(\mathbb{U}), \lambda)$ be a σ-finite measure space and $\mathbb{G} : \mathbb{U} \times \mathbb{R} \to \mathbb{R}$ a continuous mapping. Let $\mathbb{W}$ be a white noise random measure on $\mathbb{R}_+ \times \mathbb{U}$ with intensity measure λ. In this section, we consider the following BDSDE:

$$Y_t = \xi + \int_t^T \int_{\mathbb{U}} \mathbb{G}(y, Y_s)\mathbb{W}(\hat{d}sdy) - \int_t^T Z_s dB_s, \qquad 0 \le t \le T, \quad (7.1.1)$$

where ξ is an $\mathcal{F}_T^B$-measurable random variable. Recall that notation $\mathbb{W}(\hat{d}sdy)$ stands for the backward Itô integral, that is, in the Riemann sum approximating the stochastic integral, we take the right end-points instead of the left ones.

Here we regard (7.1.1) as a backward equation in two different aspects. The first is with respect to $\mathbb{W}$ for which the time variable is reversed. The second is with respect to B, which is forward in time while the boundary condition is given at the terminal instead of the initial time.

Definition 7.1.1. The pair of processes (Y_t, Z_t) is a solution to BDSDE (7.1.1) if they are $\mathcal{G}_t$-adapted, $Y_\cdot \in C([0,T], \mathbb{R})$ a.s., $\mathbb{E}\int_0^T Z_s^2 ds < \infty$, and for each $t \in [0,T]$, identity (7.1.1) holds a.s., where $\mathcal{G}_t = \sigma(\mathcal{G}_t^1, \mathcal{G}_t^2)$, $\mathcal{G}_t^1$ (non-decreasing) and $\mathcal{G}_t^2$ (non-increasing) are independent sigma-field families such that for any t, $\mathcal{F}_t^B \subset \mathcal{G}_t^1$ and $\mathcal{F}_{t,T}^{\mathbb{W}} \subset \mathcal{G}_t^2$,

$$\mathcal{F}_{t,T}^{\mathbb{W}} = \sigma\left(\mathbb{W}([r,T] \times A),\ r \in [t,T],\ A \in \mathcal{B}(\mathbb{U})\right).$$

Note that the family $\{\mathcal{G}_t\}$ of σ-fields is not a filtration because it is not increasing in time.

7.2 Itô-Pardoux-Peng formula

We now state an Itô type formula in the present setting. This formula plays a key role in the analysis of BDSDEs.

Lemma 7.2.1 (Itô-Pardoux-Peng formula). *Let y_t be the $\mathcal{G}_t$-adapted stochastic process given by*

$$y_t = \xi + \int_t^T \int_{\mathbb{U}} \alpha(s,y)\mathbb{W}(\hat{d}sdy) - \int_t^T z_s dB_s, \tag{7.2.1}$$

where $\alpha : [0,T] \times \mathbb{U} \times \Omega \to \mathbb{R}$ is a $\mathcal{G}_t$-adapted random field, and

$$\mathbb{E}\int_0^T \int_{\mathbb{U}} \alpha(s,y)^2 \lambda(dy)ds + \mathbb{E}\int_0^T z_s^2 ds < \infty.$$

Then, for any $f \in C(\mathbb{R})$ such that $f' \in C_b^1(\mathbb{R})$, we have

$$\begin{aligned} f(y_t) = f(\xi) &+ \int_t^T \int_{\mathbb{U}} f'(y_s)\alpha(s,y)\mathbb{W}(\hat{d}sdy) - \int_t^T z_s f'(y_s) dB_s \\ &+ \frac{1}{2}\int_t^T \int_{\mathbb{U}} f''(y_s)\alpha(s,y)^2 dyds - \frac{1}{2}\int_t^T z_s^2 f''(y_s)ds. \end{aligned} \tag{7.2.2}$$

Proof. Let $t_0 = t < t_1 < \cdots < t_n = T$ be a partition of $[t,T]$ such that

$$\max\{t_{i+1} - t_i : i = 0, 1, \cdots, n-1\} \to 0, \qquad \text{as } n \to \infty.$$

Then

$$\begin{aligned} f(\xi) - f(y_t) &= \sum_{i=0}^{n-1} \left(f(y_{t_{i+1}}) - f(y_{t_i})\right) \\ &= \sum_{i=0}^{n-1} \left(f'(y_{t_i})(y_{t_{i+1}} - y_{t_i}) + \frac{1}{2} f''(\theta_i)(y_{t_{i+1}} - y_{t_i})^2 \right), \end{aligned}$$

where θ_i lies between y_{t_i} and $y_{t_{i+1}}$.

Now we write out $y_{t_{i+1}} - y_{t_i}$ according to (7.2.1) as follows

$$y_{t_{i+1}} - y_{t_i} = -\int_{t_i}^{t_{i+1}} \alpha(s,y)W(\hat{d}sdy) + \int_{t_i}^{t_{i+1}} z_s dB_s,$$

and hence,

$$f(\xi) - f(y_t) = I_1 + I_2 + I_3, \tag{7.2.3}$$

where

$$I_1 = -\sum_{i=0}^{n-1} f'(y_{t_i}) \int_{t_i}^{t_{i+1}} \int_{\mathbb{U}} \alpha(s,y)\mathbb{W}(\hat{d}sdy),$$

$$I_2 = \sum_{i=0}^{n-1} f'(y_{t_i}) \int_{t_i}^{t_{i+1}} z_s dB_s,$$

and

$$I_3 = \frac{1}{2}\sum_{i=0}^{n-1} f''(\theta_i)(y_{t_{i+1}} - y_{t_i})^2.$$

From the standard Itô integral theory, we see that as $n \to \infty$,

$$I_2 \to \int_0^T f'(y_s)z_s dB_s. \tag{7.2.4}$$

Note that

$$-I_1 = I_{11} + I_{12},$$

where

$$I_{11} = \sum_{i=0}^{n-1} f'(y_{t_{i+1}}) \int_{t_i}^{t_{i+1}} \int_{\mathbb{U}} \alpha(s,y)\mathbb{W}(\hat{d}sdy)$$

and

$$I_{12} = \sum_{i=0}^{n-1} \left(f'(y_{t_i}) - f'(y_{t_{i+1}})\right) \int_{t_i}^{t_{i+1}} \int_{\mathbb{U}} \alpha(s,y)\mathbb{W}(\hat{d}sdy).$$

Analogous to the standard Itô integral theory, we have

$$I_{11} \to \int_t^T \int_{\mathbb{U}} f'(y_s)\alpha(s,y)\mathbb{W}(\hat{d}sdy). \tag{7.2.5}$$

On the other hand,

$$\begin{aligned} I_{12} &= \sum_{i=0}^{n-1} f''(\eta_i)(y_{t_i} - y_{t_{i+1}}) \int_{t_i}^{t_{i+1}} \int_{\mathbb{U}} \alpha(s,y)\mathbb{W}(\hat{d}sdy) \\ &\to \int_t^T \int_{\mathbb{U}} f''(y_s)\alpha(s,y)^2\lambda(dy)ds, \end{aligned} \tag{7.2.6}$$

where η_i lies between y_{t_i} and $y_{t_{i+1}}$.

Similarly, we can prove that

$$2I_3 \to \int_t^T \int_{\mathbb{U}} f''(y_s)\alpha(s,y)^2\lambda(dy)ds + \int_t^T f''(y_s)z_s^2 ds. \tag{7.2.7}$$

Taking $n \to \infty$ on both sides of (7.2.3), making use of (7.2.4), (7.2.5), (7.2.6) and (7.2.7), we get the Itô-Pardoux-Peng formula. □

7.3 Uniqueness of solution

Now we are ready to establish the uniqueness of the solution to BDSDE (7.1.1). We assume that $\mathbb{G} : \mathbb{U} \times \mathbb{R} \to \mathbb{R}$ satisfies the following Hölder continuity and linear growth conditions: there is a constant $K > 0$ such that for any $u_1,\ u_2,\ u \in \mathbb{R}$,

$$\int_{\mathbb{U}} |\mathbb{G}(y, u_1) - \mathbb{G}(y, u_2)|^2 \lambda(dy) \leq K|u_1 - u_2|, \tag{7.3.1}$$

and

$$\int_{\mathbb{U}} |\mathbb{G}(y, u)|^2 \lambda(dy) \leq K(1 + |u|^2). \tag{7.3.2}$$

Theorem 7.3.1. *Suppose that conditions (7.3.1) and (7.3.2) hold. Then, BDSDE (7.1.1) has at most one solution.*

Proof. Suppose that (7.1.1) has two solutions (Y_t^i, Z_t^i), $i = 1,\ 2$. Let $\{a_k\}$ be a decreasing positive sequence defined recursively by

$$a_0 = 1 \text{ and } \int_{a_k}^{a_{k-1}} z^{-1} dz = k, \qquad k \geq 1.$$

Define ψ_k to be non-negative continuous functions supported in (a_k, a_{k-1}) satisfying

$$\int_{a_k}^{a_{k-1}} \psi_k(z) dz = 1 \text{ and } \psi_k(z) \leq 2(kz)^{-1}, \qquad \forall\ z \in \mathbb{R}.$$

Let

$$\phi_k(z) = \int_0^{|z|} dy \int_0^y \psi_k(x) dx, \qquad \forall\ z \in \mathbb{R}.$$

Then,

$$\phi_k(z) \to |z| \quad \text{and} \quad |z|\phi_k''(z) \leq 2k^{-1}. \tag{7.3.3}$$

Since

$$\begin{aligned} Y_t^1 - Y_t^2 = & \int_t^T \int_{\mathbb{U}} \left(\mathbb{G}(y, Y_s^1) - \mathbb{G}(y, Y_s^2)\right) \mathbb{W}(\hat{d}sdy) \\ & - \int_t^T \left(Z_s^1 - Z_s^2\right) dB_s, \end{aligned} \tag{7.3.4}$$

by Itô-Pardoux-Peng formula, we have

$$\begin{aligned}
&\phi_k(Y_t^1 - Y_t^2) \\
&= \int_t^T \int_{\mathbb{U}} \phi_k'(Y_s^1 - Y_s^2) \left(\mathbb{G}(y, Y_s^1) - \mathbb{G}(y, Y_s^2)\right) \mathbb{W}(\hat{d}sdy) \\
&\quad - \int_t^T \phi_k'(Y_s^1 - Y_s^2) \left(Z_s^1 - Z_s^2\right) dB_s \\
&\quad + \frac{1}{2} \int_t^T \int_{\mathbb{U}} \phi_k''(Y_s^1 - Y_s^2) \left(\mathbb{G}(y, Y_s^1) - \mathbb{G}(y, Y_s^2)\right)^2 \lambda(dy)ds \\
&\quad - \frac{1}{2} \int_t^T \phi_k''(Y_s^1 - Y_s^2) \left(Z_s^1 - Z_s^2\right)^2 ds. \qquad (7.3.5)
\end{aligned}$$

Sequence ϕ_k' being bounded and

$$\mathbb{E} \int_0^T |Z_s^1 - Z_s^2|^2 ds < \infty$$

imply that the second term on the right hand side of (7.3.5) is a conditional (given $\mathbb{W}$) square integrable martingale, and hence, its expectation is 0. Moreover, by a parallel argument, the expectation of the first term is also zero. Since the last term is non-positive, by taking expectations on both sides of (7.3.5), the following estimate is attained

$$\begin{aligned}
&\mathbb{E}\phi_k(Y_t^1 - Y_t^2) \\
&\leq \mathbb{E}\frac{1}{2} \int_t^T \int_{\mathbb{U}} \phi_k''(Y_s^1 - Y_s^2) \left(\mathbb{G}(y, Y_s^1) - \mathbb{G}(y, Y_s^2)\right)^2 \lambda(dy)ds \\
&\leq K_1 \mathbb{E} \int_t^T \phi_k''(Y_s^1 - Y_s^2)|Y_s^1 - Y_s^2| ds \\
&\leq K_2 k^{-1},
\end{aligned}$$

where the second inequality follows from the Hölder continuity of $\mathbb{G}$, and the third from the property (7.3.3) of ϕ_k''. Taking $k \to \infty$ and making use of Fatou's lemma, we have

$$\mathbb{E}|Y_t^1 - Y_t^2| \leq 0.$$

Therefore, $Y_t^1 = Y_t^2$ a.s. and plugging back into (7.3.4), we can get

$$\int_t^T (Z_s^1 - Z_s^2) dB_s = 0, \qquad \text{a.s.}$$

Hence, $Z_t^1 = Z_t^2$ a.s. for a.e. t, concluding the proof. □

7.4 Historical remarks

BDSDEs and their connections with SPDEs were first studied by Pardoux and Peng (1994). The material in this chapter is taken from Xiong (2012) where we extended this connection to the case with non-Lipschitz coefficient and applied to the characterization of the SBM.

Backward stochastic differential equations (BSDEs) have been studied by many authors since the work of Bismut (1978) (linear case) and Pardoux and Peng (1990) (nonlinear case). It has many applications in various fields such as optimal control and mathematical finance.

Chapter 8

From SPDE to BSDE

In this chapter, we study the third class of SPDEs arising from the superprocesses in random environments. More specifically, we investigate the SPDE satisfied by the random "distribution" corresponding to the measure-valued process.

This chapter is organized as follows: In Section 8.1, we derive the SPDE satisfied by the distribution valued process u_t given by

$$u_t(x) = X_t((-\infty, x]), \qquad \forall\, x \in \mathbb{R},$$

where X_t is the superprocess in random environment introduced in Chapter 2. Then, in Section 8.2, we consider a class of SPDEs which takes the SPDE of Section 8.1 as a special case. For this general class of SPDEs, we prove the existence of the solutions. After that, in Section 8.3, we relate this class of SPDEs to a class of BDSDEs when the solutions are of spatial derivatives. As a consequence of the uniqueness for BDSDEs, the uniqueness of the solutions is then obtained for this restricted class of SPDEs, which includes the SPDEs for the superprocesses in random environments as special cases. Finally, for the more general class of SPDEs, whose solutions are not necessarily differentiable in the spatial variable, we mimic the uniqueness argument of the last chapter to establish their uniqueness property.

8.1 The SPDE for the distribution

This section is a generalization of Section 1.5. Instead of the SBM, we will derive the SPDE satisfied by the "distribution function"-valued process corresponding to the superprocess in a random environment.

As in Section 1.5, we define the "distribution function"-valued process as

$$u_t(x) \equiv X_t((-\infty, x]), \qquad \forall\, x \in \mathbb{R}. \tag{8.1.1}$$

Recall that

$$\langle X_t, f\rangle = -\langle u_t, f'\rangle_0 . \tag{8.1.2}$$

Then,

$$\langle X_t, Lf\rangle = -\left\langle u_t, \frac{1}{2}(af'')'\right\rangle_0 = -\left\langle \tilde{L}u_t, f'\right\rangle, \tag{8.1.3}$$

where

$$\tilde{L}u_t = \frac{1}{2}\nabla(a\nabla u_t).$$

Therefore, martingale $M_t(f)$ defined by (2.1.6) can be rewritten as

$$M_t(f) = -\langle u_t, f'\rangle_0 + \langle u_0, f'\rangle_0 + \int_0^t \left\langle \tilde{L}u_s, f'\right\rangle ds.$$

Furthermore, by substitution of variables, we get

$$\begin{aligned}
\langle M(f)\rangle_t &= \int_0^t \langle X_s, \gamma f^2\rangle\, ds + \int_0^t \int_U |\langle X_s, h(y,\cdot)f'\rangle|^2\, \mu(dy)ds \\
&= \gamma \int_0^t \int_{\mathbb{R}} f^2(x)du_s(x)ds + \int_0^t \int_U |\langle u_s, \nabla(h(y,\cdot)f')\rangle_0|^2\, \mu(dy)ds \\
&= \gamma \int_0^t \int_0^\infty f^2(u_s^{-1}(v))dvds + \int_0^t \int_U |\langle \nabla u_s, h(y,\cdot)f'\rangle|^2\, \mu(dy)ds \\
&= \gamma \int_0^t \int_0^\infty \left(\int_{\mathbb{R}} f'(x)1_{v\le u_s(x)}dx\right)^2 dvds \\
&\quad + \int_0^t \int_U |\langle h(y,\cdot)\nabla u_s, f'\rangle|^2\, \mu(dy)ds.
\end{aligned}$$

By martingale representation theorem, there exist two independent white noise random measures W^1 and W on $\mathbb{R}_+ \times \mathbb{R}_+$ with Lebesgue measure as the intensity and on $\mathbb{R}_+ \times U$ with μ as the intensity, respectively, such that

$$\begin{aligned}
M_t(f) = \sqrt{\gamma} \int_0^t \int_0^\infty \int_{\mathbb{R}} 1_{v\le u_s(x)} f'(x)dxW^1(dsdv) \\
+ \int_0^t \int_U \langle h(y,\cdot)\nabla u_s, f'\rangle\, W(dsdy).
\end{aligned} \tag{8.1.4}$$

Substituting (8.1.2)-(8.1.4) into (2.1.6), we find that random field $u_t(x)$ satisfies that for any $f \in C_0^2(\mathbb{R})$,

$$\begin{aligned}\langle u_t, f\rangle_0 = \langle \nu, f\rangle_0 &+ \int_0^t \left\langle u_s, \tilde{L}^* f\right\rangle_0 ds \\ &+\sqrt{\gamma}\int_0^t\int_0^\infty\int_{\mathbb{R}} f(x) 1_{v\le u_s(x)} dx W^1(dsdv) \\ &+\int_0^t\int_U \langle h(y,\cdot)\nabla u_s, f\rangle\, W(dsdy).\end{aligned}$$

Namely, $(u_t(x))$ is a weak solution of the following SPDE:

$$\begin{aligned}u_t(x) = \nu(x) + \int_0^t \tilde{L} u_s(x) ds &+ \sqrt{\gamma}\int_0^t\int_0^\infty 1_{v\le u_s(x)} W^1(dsdv) \\ &+\int_0^t\int_U h(y,x)\nabla u_s(x) W(dsdy). \qquad (8.1.5)\end{aligned}$$

On the other hand, if $u_t(x)$ is a solution to SPDE (8.1.5), it is easy to show that the corresponding measure-valued process determined by (8.1.1) is a solution to martingale problem (2.1.6, 2.1.7).

Thus, we have proved the following theorem.

Theorem 8.1.1. *The measure-valued process X_t is a solution to martingale problem (2.1.6, 2.1.7) if and only if random field $\{u_t(x)\}$ defined by (8.1.1) is a solution to SPDE (8.1.5).*

For the simplicity of notation, we take $Lf = \frac{1}{2}f''$ from now on. Let $\mathbb{W}$ be a white noise random measure on $\mathbb{R}_+ \times \mathbb{U}$ with intensity λ, where $(\mathbb{U}, \mathcal{B}(\mathbb{U}), \lambda)$ is a σ-finite measure space. In the rest of this chapter, we study the existence and uniqueness for the solution to the following more general SPDE: for $t \in \mathbb{R}_+$ and $y \in \mathbb{R}$,

$$u_t(y) = F(y) + \int_0^t\int_{\mathbb{U}} G(a, y, u_s)\mathbb{W}(dsda) + \int_0^t \frac{1}{2}\Delta u_s(y) ds, \qquad (8.1.6)$$

where F is a real-valued measurable function on $\mathbb{R}$, $G: \mathbb{U}\times\mathbb{R}\times\mathbb{R} \to C(\mathbb{R})$ satisfies the following conditions: there is a constant $K > 0$ such that for any u_1, u_2, $u \in C(\mathbb{R})$, $y \in \mathbb{R}$,

$$\int_{\mathbb{U}} |G(a,y,u_1) - G(a,y,u_2)|^2 \lambda(da) \le K|u_1(y) - u_2(y)|, \qquad (8.1.7)$$

and

$$\int_{\mathbb{U}} |G(a,y,u)|^2 \lambda(da) \le K(1 + |u(y)|^2). \qquad (8.1.8)$$

We first give the definition for the solution to SPDE (8.1.6). To this end, we need to introduce the following notations. For $i \in \mathbb{N} \cup \{0\}$, let $\mathcal{X}_i$ be the Hilbert space consisting of all functions f such that $f^{(k)} \in L^2(\mathbb{R}, e^{-|x|}dx)$, where $f^{(k)}$ denotes the k-th order derivative in the sense of generalized functions. We shall denote $f^{(0)} = f$ and let the Hilbert norm $\|f\|_{\mathcal{X}_i}$ be defined as

$$\|f\|_{\mathcal{X}_i}^2 \equiv \sum_{k=0}^{i} \int_{\mathbb{R}} f^{(k)}(x)^2 e^{-|x|} dx < \infty,$$

where the corresponding inner product is denoted by $\langle \cdot, \cdot \rangle_{\mathcal{X}_i}$. Let $C_0^\infty(\mathbb{R})$ be the collection of functions with compact support and derivatives of all orders.

Remark 8.1.2. We may replace $e^{-|x|}$ by 1 in the definition of $\mathcal{X}_i$. In that case, the spaces $\mathcal{X}_i$ coincide with the spaces H_i defined before. Here we added this weight function in the definition to broaden the possible applicability of the results in this chapter. For example, we may further replace $e^{-|x|}$ by $e^{-\lambda|x|}$ to obtain an increasing family of spaces $\{\mathcal{X}_i^\lambda,\ \lambda > 0\}$ and an injection between two such spaces is compact. Such a fact is useful in applications.

On the other hand, the notation complexity due to this addition is not substantial so we added this weight aiming at future applications.

Definition 8.1.3. Suppose that $F \in \mathcal{X}_0$. A continuous $\mathcal{X}_0$-valued process $\{u_t\}$ on a stochastic basis is a weak solution to SPDE (8.1.6) if there exists a white noise random measure $\mathbb{W}$ with intensity λ such that for any $t \geq 0$ and $f \in C_0^\infty(\mathbb{R})$, we have

$$\begin{aligned}\langle u_t, f\rangle = \langle F, f\rangle &+ \int_0^t \left\langle u_s, \frac{1}{2}\Delta f\right\rangle ds \\ &+ \int_0^t \int_{\mathbb{R}} \int_U G(a, y, u_s) f(y) dy \mathbb{W}(dsda), \qquad \text{a.s.}\end{aligned} \tag{8.1.9}$$

Here let $\langle f, g\rangle = \int_{\mathbb{R}} f(x)g(x)dx$ whenever the integral is well-defined.

SPDE (8.1.6) has a strong solution if for any white noise random measure $\mathbb{W}$ on stochastic basis $(\Omega, \mathcal{F}, P, \mathcal{F}_t)$, there exists a continuous $\mathcal{X}_0$-valued $\mathcal{F}_t$-adapted process $\{u_t\}$ such that (8.1.9) holds for all $f \in C_0^\infty(\mathbb{R})$.

The following theorem is the main result of this chapter whose proof will be given in Sections 8.2-8.4.

Theorem 8.1.4. *Suppose that conditions (8.1.7) and (8.1.8) hold. If* $F \in \mathcal{X}_0$*, then SPDE (8.1.6) has a strong solution* (u_t) *satisfying*

$$\mathbb{E} \sup_{0 \le t \le T} \|u_t\|_0^2 < \infty, \tag{8.1.10}$$

and any two solutions satisfying this condition will coincide.

Finally, we point out that SPDE (8.1.5) is a special case of (8.1.6) if we take $\mathbb{U} = \mathbb{R}_+ \cup U$ (assuming $\mathbb{R}_+ \cap U = \emptyset$), and define mapping $G : \mathbb{U} \times \mathbb{R} \times C(\mathbb{R}) \to \mathbb{R}$ by

$$G(a, y, u) = \begin{cases} \sqrt{\gamma} 1_{a \le u(y)} & \text{if } a \in \mathbb{R}_+, \\ h(a, u(y)) & \text{if } a \in U \end{cases}$$

and random measure $\mathbb{W}$ on $\mathbb{R}_+ \times \mathbb{U}$ by

$$\mathbb{W}(dt \times A) = W_1(dt \times (A \cap \mathbb{R}_+)) + W(dt \times (A \cap U)).$$

8.2 Existence of solution to SPDE

In this section, we consider the existence of a solution to SPDE (8.1.6). Note that the definition of the weak solution to (8.1.6) is equivalent to the following mild formulation:

$$u_t(y) = T_t F(y) + \int_0^t \int_{\mathbb{U}} \int_{\mathbb{R}} p_{t-s}(y - z) G(a, z, u_s) dz \mathbb{W}(dsda), \tag{8.2.1}$$

where T_t is the Brownian semigroup on $\mathcal{X}_0$, that is for any $f \in \mathcal{X}_0$,

$$T_t f(x) = \int_{\mathbb{R}} p_t(x - y) f(y) dy \text{ and } p_t(x) = \frac{1}{\sqrt{2\pi t}} \exp\left(-\frac{x^2}{2t}\right).$$

Note that in the equation above, we abused the notation a bit again because T_t has been used as Brownian semigroups on different spaces such as $C_b(\mathbb{R})$ and H_0.

Before constructing a solution to (8.2.1), we prove the semigroup property for family $\{T_t\}$ to be used in later sections.

Lemma 8.2.1. $\{T_t : t \ge 0\}$ *is a strongly continuous semigroup on* $\mathcal{X}_0$.

Proof. Let K_t be the function given by

$$K_t^2 = \int_{\mathbb{R}} e^{\sqrt{t}|z|} p_1(z) dz < \infty, \qquad \forall\, t \ge 0.$$

It is easy to show that for any $f \in \mathcal{X}_0$, we have

$$\|T_t f\|_{\mathcal{X}_0} \le K_t \|f\|_{\mathcal{X}_0}. \tag{8.2.2}$$

Thus, $\{T_t,\ t \geq 0\}$ is a family of bounded linear operators on $\mathcal{X}_0$. The semigroup property is not difficult to verify so we focus on this semigroup's strong continuity.

For any $f \in C_b(\mathbb{R}) \cap \mathcal{X}_0$, it follows from the dominated convergence theorem that as $t \to 0$,

$$\|T_t f - f\|^2_{\mathcal{X}_0} \leq \int_{\mathbb{R}} \left| \int_{\mathbb{R}} \left(f(x + \sqrt{t}z) - f(x) \right) p_1(z) dz \right|^2 e^{-|x|} dx \to 0.$$

In general, for $f \in \mathcal{X}_0$, we take a sequence $f_n \in C_b(\mathbb{R}) \cap \mathcal{X}_0$ such that $\|f_n - f\|_{\mathcal{X}_0} \to 0$ as $n \to \infty$. Then,

$$\|T_t f - f\|_{\mathcal{X}_0} \leq K_t \|f_n - f\|_{\mathcal{X}_0} + \|T_t f_n - f_n\|_{\mathcal{X}_0},$$

which implies $T_t f \to f$ in $\mathcal{X}_0$ as $t \to 0$. □

In addition, we define operators $T_t^{\mathbb{U}}$ on Hilbert space

$$\mathcal{X}_0 \otimes L^2(\mathbb{U}, \lambda) = L^2(\mathbb{R} \times \mathbb{U}, e^{-|x|} dx \lambda(da))$$

as

$$T_t^{\mathbb{U}} g(a, x) = \int_{\mathbb{R}} p_t(x - y) g(a, y) dy, \qquad \forall\, t \geq 0.$$

By the same argument as in the proof of Lemma 8.2.1, we have the following result.

Lemma 8.2.2. $\{T_t^{\mathbb{U}} : t \geq 0\}$ *is a strongly continuous semigroup on space* $\mathcal{X}_0 \otimes L^2(\mathbb{U}, \lambda)$. *Furthermore, for any* $g \in \mathcal{X}_0 \otimes L^2(\mathbb{U}, \lambda)$,

$$\|T_t^{\mathbb{U}} g\|_{\mathcal{X}_0 \otimes L^2(\mathbb{U},\lambda)} \leq K_t \|g\|_{\mathcal{X}_0 \otimes L^2(\mathbb{U},\lambda)}. \tag{8.2.3}$$

Now, we come back to the construction of a solution to (8.2.1). Define a sequence of approximations by: $u_t^0(y) = F(y)$ and, for $n \geq 0$,

$$u_t^{n+1}(y) = T_t F(y) + \int_0^t \int_{\mathbb{U}} \int_{\mathbb{R}} p_{t-s}(y - z) G(a, z, u_s^n) dz \mathbb{W}(dsda). \tag{8.2.4}$$

Let

$$J(x) = \int_{\mathbb{R}} e^{-|y|} \rho(x - y) dy,$$

where ρ is the mollifier given by

$$\rho(x) = K \exp\left(-1/(1 - x^2)\right) 1_{|x|<1},$$

and K is a constant such that $\int_{\mathbb{R}} \rho(x) dx = 1$. Then, for any $m \in \mathbb{Z}_+$, there are constants c_m and C_m such that

$$c_m e^{-|x|} \leq J^{(m)}(x) \leq C_m e^{-|x|}, \qquad \forall x \in \mathbb{R},$$

(cf. Mitoma (1985), (2.1)). Therefore, we may and will replace $e^{-|x|}$ by $J(x)$ in the definition of space $\mathcal{X}_i$.

Lemma 8.2.3. *For any $p \geq 1$ and $T > 0$, there exists a constant $K_1 = K_1(p,T)$ such that for any $n \geq 0$,*

$$\mathbb{E} \sup_{t \leq T} \|u_t^n\|_{\mathcal{X}_0}^{2p} \leq K_1. \tag{8.2.5}$$

Proof. We proceed by adapting the idea of Chapter 3. Smoothing out if necessary, we may and will assume that $u_t^{n+1} \in \mathcal{X}_2$. By Itô's formula, it is easy to show that, for any $f \in C_0^\infty(\mathbb{R})$,

$$\begin{aligned} \langle u_t^{n+1}, f \rangle_{\mathcal{X}_0} = \langle F, f \rangle_{\mathcal{X}_0} &+ \int_0^t \left\langle \frac{1}{2} \Delta u_s^{n+1}, f \right\rangle_{\mathcal{X}_0} ds \\ &+ \int_0^t \int_{\mathbb{R}} \int_{\mathbb{U}} G(a, y, u_s^n) f(y) J(y) dy \mathbb{W}(dsda). \end{aligned} \tag{8.2.6}$$

Applying Itô's formula to (8.2.6) gives

$$\begin{aligned} &\langle u_t^{n+1}, f \rangle_{\mathcal{X}_0}^2 \\ &= \langle F, f \rangle_{\mathcal{X}_0}^2 + \int_0^t \langle u_s^{n+1}, f \rangle_{\mathcal{X}_0} \langle \Delta u_s^{n+1}, f \rangle_{\mathcal{X}_0} ds \\ &\quad + \int_0^t \int_{\mathbb{U}} \left(\int_{\mathbb{R}} G(a, y, u_s^n) f(y) J(y) dy \right)^2 \lambda(da) ds \\ &\quad + \int_0^t \int_{\mathbb{U}} 2 \langle u_s^{n+1}, f \rangle_{\mathcal{X}_0} \int_{\mathbb{R}} G(a, y, u_s^n) f(y) J(y) dy \mathbb{W}(dsda). \end{aligned}$$

Summing on f over a CONS of $\mathcal{X}_0$, we have

$$\begin{aligned} \|u_t^{n+1}\|_{\mathcal{X}_0}^2 = \|F\|_{\mathcal{X}_0}^2 &+ \int_0^t \langle u_s^{n+1}, \Delta u_s^{n+1} \rangle_{\mathcal{X}_0} ds \\ &+ \int_0^t \int_{\mathbb{U}} \int_{\mathbb{R}} G(a, y, u_s^n)^2 J(y) dy \lambda(da) ds \\ &+ \int_0^t \int_{\mathbb{U}} 2 \langle u_s^{n+1}, G(a, y, u_s^n) \rangle_{\mathcal{X}_0} \mathbb{W}(dsda). \end{aligned}$$

Itô's formula is again applied to obtain

$$\begin{aligned}
&\|u_t^{n+1}\|_{\mathcal{X}_0}^{2p} \\
&= \|F\|_{\mathcal{X}_0}^{2p} + \int_0^t p\|u_s^{n+1}\|_{\mathcal{X}_0}^{2(p-1)} \left\langle u_s^{n+1}, \Delta u_s^{n+1}\right\rangle_{\mathcal{X}_0} ds \\
&\quad + \int_0^t p\|u_s^{n+1}\|_{\mathcal{X}_0}^{2(p-1)} \int_{\mathbb{U}}\int_{\mathbb{R}} G(a,y,u_s^n)^2 J(y)dy\lambda(da)ds \\
&\quad + \int_0^t p\|u_s^{n+1}\|_{\mathcal{X}_0}^{2(p-1)} \int_{\mathbb{U}} 2\left\langle u_s^{n+1}, G(a,y,u_s^n)\right\rangle_{\mathcal{X}_0} \mathbb{W}(dsda) \\
&\quad + 2p(p-1)\int_0^t \|u_s^{n+1}\|_{\mathcal{X}_0}^{2(p-2)} \int_{\mathbb{U}} \left\langle u_s^{n+1}, G(a,y,u_s^n)\right\rangle_{\mathcal{X}_0}^2 \lambda(da)ds.
\end{aligned} \tag{8.2.7}$$

Note that, for $u \in \mathcal{X}_1$,

$$\int_{\mathbb{R}} u(x)u'(x)J'(x)dx = -\int_{\mathbb{R}} u(x)(u'(x)J'(x) + u(x)J''(x))dx,$$

which implies that

$$\begin{aligned}
-\int_{\mathbb{R}} u(x)u'(x)J'(x)dx &= \frac{1}{2}\int_{\mathbb{R}} u(x)^2 J''(x)dx \\
&\le K_2 \int_{\mathbb{R}} u(x)^2 J(x)dx \\
&= K_2\|u\|_{\mathcal{X}_0}^2.
\end{aligned}$$

Therefore,

$$\begin{aligned}
\langle u, \Delta u\rangle_{\mathcal{X}_0} &= \int_{\mathbb{R}} u''(x)u(x)J(x)dx \\
&= -\int_{\mathbb{R}} u'(x)(u'(x)J(x) + u(x)J'(x))dx \\
&\le K_2\|u\|_{\mathcal{X}_0}^2.
\end{aligned}$$

By applying Burkholder-Davis-Gundy inequality to (8.2.7), we obtain

$$\begin{aligned}
\mathbb{E}\sup_{s\le t}\|u_s^{n+1}\|_{\mathcal{X}_0}^{2p} &\le \|F\|_{\mathcal{X}_0}^{2p} + pK_2\int_0^t \mathbb{E}\|u_s^{n+1}\|_{\mathcal{X}_0}^{2p} ds \\
&\quad + K_3\int_0^t \mathbb{E}\left(\|u_s^{n+1}\|_{\mathcal{X}_0}^{2(p-1)}\left(1 + \|u_s^n\|_{\mathcal{X}_0}^2\right)\right)ds \\
&\quad + K_4\mathbb{E}\left(\int_0^t \|u_s^{n+1}\|_{\mathcal{X}_0}^{4p-2}\left(1 + \|u_s^n\|_{\mathcal{X}_0}^2\right)ds\right)^{1/2}.
\end{aligned}$$

Hence,

$$\begin{aligned}
f_{n+1}(t) &\equiv \mathbb{E}\sup_{s\le t}\|u_s^{n+1}\|_{\mathcal{X}_0}^{2p} \\
&\le \|F\|_{\mathcal{X}_0}^{2p} + K_5\int_0^t f_{n+1}(s)ds + K_6\int_0^t f_n(s)ds + \frac{1}{2}f_{n+1}(t).
\end{aligned}$$

The Gronwall's inequality and an induction argument finish the proof. □

We proceed to proving the tightness of $\{u^n\}$ in $C([0,T]\times\mathbb{R})$. Denote

$$v_t^n(y)=\int_0^t\int_{\mathbb{U}}\int_{\mathbb{R}}p_{t-s}(y-z)G(a,z,u_s^n)dz\mathbb{W}(dsda).$$

Lemma 8.2.4. *For any $p\ge 1>\alpha$, there is a constant K_1 such that*

$$\mathbb{E}|v_t^n(y_1)-v_t^n(y_2)|^{2p}\le K_1e^{p(|y_1|\vee|y_2|)}|y_1-y_2|^{p\alpha}. \tag{8.2.8}$$

Proof. Denote the left hand side of (8.2.8) by I. It follows from Burkholder's inequality that there exists a constant $K_2>0$ such that I is bounded by

$$K_2\mathbb{E}\left(\int_0^t\int_{\mathbb{U}}\left(\int_{\mathbb{R}}(p_s(y_1-z)-p_s(y_2-z))\,G(a,z,u_{t-s}^n)dz\right)^2\lambda(da)ds\right)^p.$$

By Hölder's inequality,

$$\begin{aligned}I\le K_2\mathbb{E}\bigg(\int_0^t\int_{\mathbb{U}}\int_{\mathbb{R}}&(p_s(y_1-z)-p_s(y_2-z))^2e^{|z|}dz\\&\times\int_{\mathbb{R}}G(a,z,u_{t-s}^n)^2e^{-|z|}dz\lambda(da)ds\bigg)^p.\end{aligned}$$

The linear growth condition (8.1.8) and estimate (8.2.5) is then applied to get

$$\begin{aligned}I&\le K_2\mathbb{E}\bigg(\int_0^t\int_{\mathbb{R}}(p_s(y_1-z)-p_s(y_2-z))^2e^{|z|}dz\\&\qquad\times\int_{\mathbb{R}}K(1+|u_{t-s}^n(z)|^2)e^{-|z|}dzds\bigg)^p\\&\le K_3\left(\int_0^t\int_{\mathbb{R}}(p_s(y_1-z)-p_s(y_2-z))^2e^{|z|}dzds\right)^p.\end{aligned}$$

Using the fact that

$$|p_s(y_1)-p_s(y_2)|\le Ks^{-1}|y_1-y_2|,\qquad\forall\, s>0,\;\; y_1,\;y_2\in\mathbb{R},$$

we arrive at

$$\begin{aligned}I&\le K_4\left(\int_0^t\int_{\mathbb{R}}s^{-\alpha}|y_1-y_2|^\alpha\,(p_s(y_1-z)\vee p_s(y_2-z))^{2-\alpha}e^{|z|}dzds\right)^p\\&\le K_4\left(\int_0^t\int_{\mathbb{R}}s^{-\alpha}|y_1-y_2|^\alpha p_s(z)^{2-\alpha}e^{|z|}dzdse^{|y_1|\vee|y_2|}\right)^p\\&\le K_5\left(\int_0^t s^{-\alpha}s^{-(1-\alpha)/2}ds\right)^p e^{p(|y_1|\vee|y_2|)}|y_1-y_2|^{p\alpha}\\&\le K_1e^{p(|y_1|\vee|y_2|)}|y_1-y_2|^{p\alpha},\end{aligned}$$

which finishes the proof of (8.2.8). □

Similarly, we can prove that

$$\mathbb{E}|v^n_{t_1}(y) - v^n_{t_2}(y)|^{2p} \le K_1 e^{p|y|/2}|t_1 - t_2|^{p\alpha/2}.$$

We are now ready to provide *the proof of Theorem 8.1.4 (Existence).*

Proof. It follows from Kolmogorov's criteria (see Corollary 16.9 in Kallenberg (2002)) that, for each fixed m, the sequence of laws of

$$\{v^n_t(x) : \ (t,x) \in [0,T] \times [-m,m]\}$$

on $C([0,T]\times[-m,m])$ is tight, and hence, has a convergent subsequence. By the standard diagonalization argument, there exists a subsequence $\{v^{n_k}_t(x)\}$ which converges in law on $C([0,T] \times [-m,m])$ for each m. Therefore, $\{v^{n_k}_t(x)\}$ converges in law on $C([0,T] \times \mathbb{R})$.

Let $v_t(x)$ be a limit point. For any $t_1 < t_2$, it follows from Fatou's lemma that

$$\begin{aligned}
\mathbb{E}\|v_{t_1} - v_{t_2}\|^{2p}_{\mathcal{X}_0} &\le K_1 \liminf_{k\to\infty} \mathbb{E}\left(\int_{\mathbb{R}} |v^{n_k}_{t_1}(x) - v^{n_k}_{t_2}(x)|^2 e^{-|x|}dx\right)^p \\
&\le K_2 \liminf_{k\to\infty} \mathbb{E}\int_{\mathbb{R}} |v^{n_k}_{t_1}(x) - v^{n_k}_{t_2}(x)|^{2p} e^{-\frac{2}{3}p|x|}dx \\
&\le K_3 \int_{\mathbb{R}} e^{\frac{1}{2}p|x|}|t_1 - t_2|^{p\alpha/2} e^{-\frac{2}{3}p|x|}dx \\
&= K_4|t_1 - t_2|^{p\alpha/2}.
\end{aligned}$$

By Kolmogorov's criteria again, we see that there is a version, which we will take, such that $v. \in C([0,T], \mathcal{X}_0)$ a.s.

Let $u_t(y) = T_tF(y) + v_t(y)$. Then, $u. \in C([0,T], \mathcal{X}_0)$ a.s. The proof of $\{u.\}$ being a solution to SPDE (8.1.6) is standard and is similar to what we did a few times in previous chapters so we only give a sketch here. First, by passing to the limit, we can prove that for any $f \in C^\infty_0(\mathbb{R})$,

$$M^f_t \equiv \langle u_t, f\rangle - \langle F, f\rangle - \int_0^t \left\langle u_s, \frac{1}{2}\Delta f\right\rangle ds$$

and

$$\begin{aligned}
N^f_t \equiv \langle u_t, f\rangle^2 - \langle F, f\rangle^2 &- \int_0^t \langle u_s, f\rangle \langle u_s, \Delta f\rangle ds \\
&- \int_0^t \int_{\mathbb{U}} \left(\int_{\mathbb{R}} G(a,y,u_s)f(y)dy\right)^2 \lambda(da)ds
\end{aligned}$$

are martingales. It then follows that the quadratic variation process of M^f is given by

$$\langle M^f\rangle_t = \int_0^t \int_{\mathbb{U}} \left(\int_{\mathbb{R}} G(a,y,u_s)f(y)dy\right)^2 \lambda(da)ds.$$

Martingale M^f is then represented as

$$M_t^f = \int_0^t \int_{\mathbb{R}} \int_{\mathbb{U}} G(a, y, u_s) f(y) dy \mathbb{W}(dsda)$$

for a suitable random measure $\mathbb{W}$ defined on a stochastic basis. Consequently, u_t is a weak solution to SPDE (8.1.6).

Estimate (8.1.10) follows from (8.2.5) and Fatou's lemma. □

8.3 From BSDE to SPDE

Finally, in this section, we establish a relationship between SPDEs and BDSDEs under the non-Lipschitz setup. To this end, we convert SPDE (8.1.6) to its backward version. For T fixed, we define random field

$$\tilde{u}_t(y) = u_{T-t}(y), \qquad \forall\, t \in [0, T], \;\; y \in \mathbb{R},$$

and introduce a new noise $\tilde{\mathbb{W}}$ by

$$\tilde{\mathbb{W}}([0, t] \times A) = \mathbb{W}([T - t, T] \times A), \qquad \forall\, t \in [0, T], \;\; A \in \mathcal{B}(\mathbb{U}).$$

Then, $\tilde{u}_t$ satisfies the backward SPDE given by

$$\tilde{u}_t(y) = F(y) + \int_t^T \int_{\mathbb{U}} G(a, y, \tilde{u}_s) \tilde{\mathbb{W}}(\hat{d}sda) + \int_t^T \frac{1}{2} \Delta \tilde{u}_s(y) ds. \tag{8.3.1}$$

It is clear that SPDEs (8.1.6) and (8.3.1) have the same uniqueness property. Specifically, if (8.1.6) has a unique strong solution, then so does (8.3.1), and vice versa. Observe that $\tilde{u}_t$ is $\mathcal{F}_{t,T}^{\tilde{\mathbb{W}}}$-measurable.

Now, we suppose there exists a function $\mathbb{G} : \mathbb{U} \times \mathbb{R} \to \mathbb{R}$ such that

$$G(a, y, u) = \mathbb{G}(a, u(y)), \qquad \forall\, a \in \mathbb{U}, \; y \in \mathbb{R}, \; u \in C(\mathbb{R}).$$

We denote

$$X_s^{t,y} = y + B_s - B_t, \qquad \forall\, t \le s \le T, \tag{8.3.2}$$

and consider the following BDSDE: For $t \le s \le T$,

$$Y_s^{t,y} = F(X_T^{t,y}) + \int_s^T \int_{\mathbb{U}} \mathbb{G}(a, Y_r^{t,y}) \tilde{\mathbb{W}}(\hat{d}rda) - \int_s^T Z_r^{t,y} dB_r. \tag{8.3.3}$$

BDSDE (8.3.3) coincides with BDSDE (7.1.1) if we fix (t, y), take $\xi = F(X_T^{t,y})$, and let the initial time be denoted by t instead of 0 (t is fixed and s varies as shown). We use the superscript (t, y) to indicate the dependency on the initial state of the underlying motion.

Theorem 8.3.1. *Suppose that conditions (8.1.7) and (8.1.8) hold. If process* $\{\tilde{u}_t\}$ *is a solution to (8.3.1) such that* $\tilde{u}_\cdot \in C([0,T],\mathcal{X}_1)$ *a.s., and*

$$\mathbb{E}\int_0^T \|\tilde{u}_s\|_1^2 ds < \infty, \tag{8.3.4}$$

then

$$\tilde{u}_t(y) = Y_t^{t,y},$$

where $Y_s^{t,y}$ *is a solution to BDSDE (8.3.3).*

Proof. Let

$$Y_s^{t,y} = \tilde{u}_s(X_s^{t,y}) \text{ and } Z_s^{t,y} = \nabla\tilde{u}_s(X_s^{t,y}), \qquad t \le s \le T. \tag{8.3.5}$$

To prove (8.3.3), we need to smooth the function $\tilde{u}_t$. For any $\delta > 0$, let

$$u_t^\delta(y) = T_\delta \tilde{u}_t(y), \qquad \forall\, y \in \mathbb{R}.$$

It is well-known that for any $t \ge 0$ and $\delta > 0$, $u_t^\delta \in C^\infty$. Applying T_δ to both sides of (8.3.1), we have

$$\begin{aligned} u_t^\delta(y) = T_\delta F(y) + \int_t^T \frac{1}{2}\Delta u_s^\delta(y) ds \\ + \int_t^T \int_{\mathbb{U}}\int_{\mathbb{R}} p_\delta(y-z) G(a,z,\tilde{u}_s) dz \tilde{\mathbb{W}}(\hat{d}sda). \end{aligned} \tag{8.3.6}$$

Let $s = t_0 < t_1 < \cdots < t_n = T$ be a partition of $[s,T]$. Then,

$$\begin{aligned} & u_s^\delta(X_s^{t,y}) - T_\delta F(X_T^{t,y}) \\ &= \sum_{i=0}^{n-1}\left(u_{t_i}^\delta(X_{t_i}^{t,y}) - u_{t_i}^\delta(X_{t_{i+1}}^{t,y})\right) + \sum_{i=0}^{n-1}\left(u_{t_i}^\delta(X_{t_{i+1}}^{t,y}) - u_{t_{i+1}}^\delta(X_{t_{i+1}}^{t,y})\right) \\ &= -\sum_{i=0}^{n-1}\int_{t_i}^{t_{i+1}} \frac{1}{2}\Delta u_{t_i}^\delta(X_r^{t,y}) dr - \sum_{i=0}^{n-1}\int_{t_i}^{t_{i+1}} \nabla u_{t_i}^\delta(X_r^{t,y}) dB_r \\ &\quad + \sum_{i=0}^{n-1}\int_{t_i}^{t_{i+1}} \frac{1}{2}\Delta u_r^\delta(X_{t_{i+1}}^{t,y}) dr \\ &\quad + \sum_{i=0}^{n-1}\int_{t_i}^{t_{i+1}} \int_{\mathbb{R}}\int_{\mathbb{U}} p_\delta(X_{t_{i+1}}^{t,y} - z) G(a,z,\tilde{u}_r) \tilde{\mathbb{W}}(\hat{d}rda) dz, \end{aligned}$$

where we used Itô's formula for $u_{t_i}^\delta$ (note that $u_{t_i}^\delta$ is independent of $X_r^{t,y}$ and B_r), and SPDE (8.3.6) with y replaced by $X_{t_{i+1}}^{t,y}$. Setting the mesh size to go to 0, we obtain

$$\begin{aligned} & u_s^\delta(X_s^{t,y}) - T_\delta F(X_T^{t,y}) \\ = & -\int_s^T \nabla u_r^\delta(X_r^{t,y})dB_r \\ & +\int_s^T \int_{\mathbb{R}} \int_{\mathbb{U}} p_\delta(X_r^{t,y} - z)G(a,z,\tilde{u}_r)\tilde{\mathbb{W}}(\hat{d}rda)dz. \end{aligned} \tag{8.3.7}$$

We take $\delta \to 0$ on both sides of (8.3.7). Note that for $s > t$,

$$\begin{aligned} & \mathbb{E}\left|\int_s^T \nabla u_r^\delta(X_r^{t,y})dB_r - \int_s^T \nabla\tilde{u}_r(X_r^{t,y})dB_r\right|^2 \\ = & \mathbb{E}\int_s^T \left|\nabla u_r^\delta(X_r^{t,y}) - \nabla\tilde{u}_r(X_r^{t,y})\right|^2 dr \\ = & \mathbb{E}\int_s^T \int_{\mathbb{R}} (T_\delta\nabla\tilde{u}_r(z) - \nabla\tilde{u}_r(z))^2 p_{r-t}(y-z)dzdr. \end{aligned}$$

For $s > t$ fixed, there exists a constant K_1, depending on $s - t$, such that for any $r > s$,

$$p_{r-t}(y-z) \le K_1 e^{-|y-z|} \le K_1 e^{|y|}e^{-|z|}.$$

Thus, we may continue the estimate above with

$$\begin{aligned} & \mathbb{E}\left|\int_s^T \nabla u_r^\delta(X_r^{t,y})dB_r - \int_s^T \nabla\tilde{u}_r(X_r^{t,y})dB_r\right|^2 \\ \le & K_1 e^{|y|}\mathbb{E}\int_s^T \int_{\mathbb{R}} (T_\delta\nabla\tilde{u}_r(z) - \nabla\tilde{u}_r(z))^2 e^{-|z|}dzdr \to 0, \end{aligned}$$

where the last step follows from the integrability condition (8.3.4).

Other terms can be estimated similarly. (8.3.3) follows from (8.3.7) by taking $\delta \to 0$. □

8.4 Uniqueness for SPDE

The existence of a solution to SPDE (8.1.6) was established in Section 8.2. This section is devoted to the proof of the uniqueness part of Theorem 8.1.4.

Proof. Let u_s^j, $j = 1, 2$, be two solutions to SPDE (8.1.6). Let $T > 0$ be fixed and let $\tilde{u}_s^j = u_{T-s}^j$. Denote

$$u_s^{j,\delta} = T_\delta\tilde{u}_s^j, \qquad \text{for } j = 1,\ 2$$

and let $s > t$ be fixed. By (8.3.7),

$$\begin{aligned} & u_s^{1,\delta}(X_s^{t,y}) - u_s^{2,\delta}(X_s^{t,y}) \qquad (8.4.1) \\ = & -\int_s^T \nabla \left(u_s^{1,\delta} - u_s^{2,\delta}\right)(X_r^{t,y})dB_r \\ & + \int_s^T \int_{\mathbb{U}} \int_{\mathbb{R}} p_\delta(X_r^{t,y} - z)\left(G(a,z,\tilde{u}_r^1) - G(a,z,\tilde{u}_r^2)\right) dz\tilde{\mathbb{W}}(\hat{d}rda). \end{aligned}$$

Let ϕ_k be defined as in the proof of Theorem 7.3.1. Applying Itô-Pardoux-Peng formula to (8.4.1) and ϕ_k, similar to (7.3.5), we get

$$\begin{aligned} & \mathbb{E}\phi_k\left(u_s^{1,\delta}(X_s^{t,y}) - u_s^{2,\delta}(X_s^{t,y})\right) \qquad (8.4.2) \\ \le & \frac{1}{2}\mathbb{E}\int_s^T \int_{\mathbb{U}} \phi_k''\left(u_r^{1,\delta}(X_r^{t,y}) - u_r^{2,\delta}(X_r^{t,y})\right) \\ & \left|\int_{\mathbb{R}} p_\delta(X_r^{t,y} - z)\left(G(a,z,\tilde{u}_r^1) - G(a,z,\tilde{u}_r^2)\right) dz\right|^2 \lambda(da)dr. \end{aligned}$$

Next, we take the limit $\delta \to 0$ on both sides of (8.4.2). By Lemma 8.2.1, $T_\delta \tilde{u}_s^j \to \tilde{u}_s^j$ in $\mathcal{X}_0$ as $\delta \to 0$. Taking a subsequence if necessary, we may and will assume that $T_\delta \tilde{u}_s^j(x) \to \tilde{u}_s^j(x)$ for almost every x with respect to the Lebesgue measure. Therefore,

$$u_s^{1,\delta}(X_s^{t,y}) - u_s^{2,\delta}(X_s^{t,y}) \to \tilde{u}_s^1(X_s^{t,y}) - \tilde{u}_s^2(X_s^{t,y}), \qquad \text{a.s.,}$$

and by the bounded convergence theorem, the left hand side of (8.4.2) converges to

$$\mathbb{E}\phi_k\left(\tilde{u}_s^1(X_s^{t,y}) - \tilde{u}_s^2(X_s^{t,y})\right).$$

Now we denote

$$g_r(a,z) = G(a,z,\tilde{u}_r^1) - G(a,z,\tilde{u}_r^2), \qquad (a,z) \in \mathbb{U} \times \mathbb{R}.$$

Then, the right hand side of (8.4.2) can be written as

$$\begin{aligned} & \frac{1}{2}\mathbb{E}\int_s^T \int_{\mathbb{R}} \int_{\mathbb{U}} \phi_k''\left(u_r^{1,\delta}(x) - u_r^{2,\delta}(x)\right)|T_\delta^{\mathbb{U}} g_r(a,x)|^2 p_{r-t}(x-y)dx\lambda(da)dr \\ = & \frac{1}{2}\mathbb{E}\int_s^T \|(T_\delta^{\mathbb{U}} g_r)h_r^\delta\|^2_{\mathcal{X}_0 \otimes L^2(\mathbb{U},\lambda)} dr, \qquad (8.4.3) \end{aligned}$$

where $h_r^\delta(x)$, $r \ge s$ and $x \in \mathbb{R}$, is such that

$$h_r^\delta(x)^2 = \phi_k''(u_r^{1,\delta}(x) - u_r^{2,\delta}(x))e^{|x|}p_{r-t}(x-y).$$

We note that $h_r(x)$ is bounded by a constant depending on $(k,\ s-t,\ y)$. On the other hand,

$$\|g_r\|^2_{\mathcal{X}_0 \otimes L^2(\mathbb{U},\lambda)} \le K \int_{\mathbb{R}} \left(1 + |u_r^1(z)|^2 + |u_r^2(z)|^2\right) e^{-|z|}dz$$

which is integrable. By Lemma 8.2.2 and the dominated convergence theorem, we see that the limit of the right hand side of (8.4.2) as $\delta \to 0$ is equal to

$$\begin{aligned}
&\frac{1}{2}\mathbb{E}\int_s^T \lim_{\delta\to 0} \|T_\delta^{\mathbb{U}} g_r h_r^\delta\|^2_{\mathcal{X}_0\otimes L^2(\mathbb{U},\lambda)} dr \\
=&\frac{1}{2}\mathbb{E}\int_s^T \|g_r h_r\|^2_{\mathcal{X}_0\otimes L^2(\mathbb{U},\lambda)} dr \\
\le& K_1\mathbb{E}\int_s^T \phi_k''\left(\tilde{u}_r^1(X_r^{t,y}) - \tilde{u}_r^2(X_r^{t,y})\right)|\tilde{u}_r^1(X_r^{t,y}) - \tilde{u}_r^2(X_r^{t,y})|dr,
\end{aligned}$$

where K_1 is a constant and h_r is defined as

$$h_r(x)^2 = \phi_k''(u_r^1(x) - u_r^2(x))e^{|x|}p_{r-t}(x-y).$$

To summarize, we obtain

$$\begin{aligned}
&\mathbb{E}\phi_k\left(\tilde{u}_s^1(X_s^{t,y}) - \tilde{u}_s^2(X_s^{t,y})\right) \qquad (8.4.4)\\
\le& K_1\mathbb{E}\int_s^T \phi_k''\left(\tilde{u}_r^1(X_r^{t,y}) - \tilde{u}_r^2(X_r^{t,y})\right)|\tilde{u}_r^1(X_r^{t,y}) - \tilde{u}_r^2(X_r^{t,y})|dr \\
\le& k^{-1}TK_1,
\end{aligned}$$

where we used $|z|\phi_k''(z) \le 2k^{-1}$ in the last step.

Finally, applying Fatou's lemma for $k \to \infty$, we obtain,

$$\mathbb{E}|\tilde{u}_s^1(X_s^{t,y}) - \tilde{u}_s^2(X_s^{t,y})| \le \liminf_{k\to\infty}\mathbb{E}\phi_k\left(\tilde{u}_s^1(X_s^{t,y}) - \tilde{u}_s^2(X_s^{t,y})\right) \le 0.$$

Therefore,

$$\tilde{u}_s^1(X_s^{t,y}) - \tilde{u}_s^2(X_s^{t,y}) = 0, \qquad \text{a.s.}$$

Taking $s \downarrow t$, we conclude

$$u_t^1(y) = u_t^2(y), \qquad \text{a.s.}$$

□

After proving the pathwise (strong) uniqueness and weak existence of the solution for SPDE (8.1.6), we verify its (weak) uniqueness. To apply Kurtz's result (see Theorem A.3.3) to SPDE (8.1.6), we convert it to an SPDE driven by a sequence of independent Brownian motions. Let $\{h_j\}_{j=1}^\infty$ be a CONS of $L^2(\mathbb{U}, \mathcal{B}(\mathbb{U}), \lambda)$ and define

$$B_t^j = \int_0^t\int_{\mathbb{U}} h_j(a)\mathbb{W}(dsda), \qquad j = 1, 2, \cdots.$$

Letting $B_t = (B_t^j)_{j=1}^\infty$, it is easy to see that (8.1.6) is equivalent to the following SPDE

$$u_t(y) = F(y) + \sum_{j=1}^\infty \int_0^t G_j(y, u_s) dB_s^j + \int_0^t \frac{1}{2} \Delta u_s(y) ds, \tag{8.4.5}$$

where

$$G_j(y, u) = \int_{\mathbb{U}} G(a, y, u) h_j(a) \lambda(da).$$

Denote

$$S_1 = C([0,T], \mathcal{X}_0) \text{ and } S_2 = C([0,T], \mathbb{R}^\infty).$$

Let $\{f_k\}_{k=1}^\infty \subset C_0^\infty(\mathbb{R})$ be a dense subset of $\mathcal{X}_0$ and $\Gamma : S_1 \times S_2 \to \mathbb{R}$ be the measurable functional defined by

$$\Gamma(u_\cdot, B_\cdot) = \sum_{k=1}^\infty \sup_{t \le T} |\gamma_t^k| \wedge 2^{-k},$$

where

$$\Gamma_t^k = \langle u_t, f_k \rangle - \langle F, f_k \rangle - \int_0^t \left\langle u_s, \frac{1}{2} \Delta f_k \right\rangle ds$$
$$- \sum_{j=1}^\infty \int_0^t \int_{\mathbb{R}} G_j(y, u_s) f(y) dy dB_s^j.$$

Then, SPDE (8.4.5) can be rewritten as

$$\Gamma(u_\cdot, B_\cdot) = 0.$$

The following theorem is a direct consequence of Theorem A.3.3.

Theorem 8.4.1. *If $(u_\cdot^i)$, $i = 1, 2$, are two solutions of SPDE (8.1.6) (may be defined on different stochastic bases) such that*

$$\mathbb{E} \sup_{t \le T} \|u_t^i\|_0^2 < \infty, \qquad i = 1, 2,$$

then, their laws in $C([0,T], \mathcal{X}_0)$ coincide.

8.5 Historical remarks

This chapter is based on the paper of Xiong (2012) which is inspired by Dawson and Li (2012) in which they considered SPDE with $\frac{1}{2}\Delta$ replaced by the bounded operator A given by

$$Af(x) = (\gamma(x) - f(x))b,$$

where b is a constant and γ is a fixed function.

We conclude this chapter by mentioning other possible applications of the idea developed in this chapter. The first is to consider measure-valued processes with interaction among individuals in the system. This interaction may come from the drift and diffusion coefficients which govern the motion of the individuals. It may also come from the branching and immigration mechanisms. This extension will appear in a joint work of Mytnik and Xiong (2012). The second possible application is to consider other types of nonlinear SPDEs, especially those where the noise term involves the spatial derivative of the solution. This extension will appear in a joint work of Gomez, Lee, Mueller, Wei and Xiong (2012). Finally, studying measure-valued processes by using SPDE methodology has the advantage of utilizing the rich collection of tools developed in the area of SPDEs. For example, the large deviation principle (LDP) for some measure-valued processes, including Fleming-Viot process and the SBM, can be established. As it is well-known, LDP for general FV process is a long standing open problem (some partial results were obtained by Dawson and Feng (1998), (2001), and Feng and Xiong (2002) for neutral FV processes, and Xiang and Zhang (2005) for the case when the mutation operator tends to 0). This application will be presented in a joint work of Fatheddin and Xiong (2012).

Finally, we would like to point out that there are other non-standard SPDEs related to SPREs studied in the literature. For example, Li, Wang and Xiong (2004) studied a degenerated SPDE obtained as the scaling limit of the SPRE; Li and Ma (2008) characterize a SPRE as the unique solution to an SPDE driven by a Poisson random measure.

Appendix

Some Auxiliary Results

A.1 Martingale representation theorems

In this appendix, we introduce the basic definitions about the white noise random measure (WNRM) and the martingale representation theorem with respect to such random measures. We refer the reader to the book of Kallianpur and Xiong (1995) for more details.

Definition A.1.1. Let $(U,\mathcal{U})$ be a measurable space and let $(\Omega,\mathcal{F},P,\mathcal{F}_t)$ be a stochastic basis satisfying the usual conditions. A mapping $W:\Omega\times(\mathcal{B}(\mathbb{R}_+)\times\mathcal{U})\to\mathbb{R}$ is called a *random measure* if $W(\omega,\cdot)$ is a signed measure on $\mathbb{R}_+\times U$ for each ω and $W(\cdot,B)$ is a random variable for each $B\in\mathcal{B}(\mathbb{R}_+)\times\mathcal{U}$. A random measure W is called *adapted* if $W(\cdot,B)$ is $\mathcal{F}_t$-measurable for $B\subset[0,t]\times U$.

A random measure W is *σ-finite* if there exists a sequence U_n increasing to U such that $\mathbb{E}|W(\cdot,[0,t]\times U_n)|<\infty$ for each $n\geq 1$ and $t>0$.

A random measure W is called a *martingale random measure* if for any $A\in\Gamma_W$, stochastic process $\xi_t\equiv W([0,t]\times A)$ is a martingale, where

$$\Gamma_W\equiv\{A\in\mathcal{U}:\ \mathbb{E}|W([0,t]\times A)|<\infty\ \forall t>0\}.$$

Definition A.1.2. A random measure W is called *independently scattered* if for any disjoint $B_1,\cdots,B_n\in\mathcal{B}(\mathbb{R}_+)\times\mathcal{U}$, random variables $W(\cdot,B_1),\cdots,W(\cdot,B_n)$ are independent.

An independently scattered adapted random measure is called a *white noise random measure* if there exists a σ-finite measure μ on $(U,\mathcal{U})$ such that for any $B\in\mathcal{B}(\mathbb{R}_+)\times\mathcal{U}$ with $(dtd\mu)(B)<\infty$, $W(\cdot,B)$ is a Gaussian random variable with mean 0 and variance $(dtd\mu)(B)$. μ is called the *intensity measure* of W.

Now, we assume that W is a white noise random measure on $\mathbb{R}_+ \times U$ with intensity μ. For any simple function ϕ on U given by

$$\phi(y) = \sum_{j=1}^{n} c_j 1_{A_j}(y),$$

we define the stochastic integral as

$$W_t(\phi) = \sum_{j=1}^{n} c_j W([0,t] \times A_j).$$

It is easy to show that $W_t(\phi)$ is a continuous martingale with quadratic variation process

$$\langle W(\phi)\rangle_t = \|\phi\|^2_{L^2(U,\mu)} t. \tag{A.1.1}$$

For any $\phi \in L^2(U,\mu)$, we take (ϕ^n) to be a sequence of simple functions and converges to ϕ in $L^2(U,\mu)$. The Burkholder-Davis-Gundy inequality implies that

$$\mathbb{E} \sup_{0\le t\le T} |W_t(\phi^n) - W_t(\phi^m)|^2 \le 4\|\phi^n - \phi^m\|^2_{L^2(U,\mu)} T \to 0$$

as $m,\ n \to \infty$. Namely, $\{W(\phi^n)\}$ is a Cauchy sequence. Denote the limit by $W(\phi)$. It is then easy to show that $W_t(\phi)$ is a continuous martingale with quadratic variation process given by (A.1.1).

Let $\{\phi^j\}_{j=1}^{\infty}$ be a CONS of Hilbert space $L^2(U,\mu)$. Define

$$W_t^j = \int_0^t \int_U \phi^j(y) W(dsdy), \quad j = 1, 2, \cdots.$$

Then, $\{W^j\}_{j=1}^{\infty}$ are independent Brownian motions.

Definition A.1.3. A mapping $f:\ \mathbb{R}_+ \times \Omega \to L^2(U,\mu)$ is *predictable* if for any $\phi \in L^2(U,\mu)$, the real-valued process $(t,\omega) \mapsto \langle f(t,\omega), \phi\rangle_{L^2(U,\mu)}$ is predictable.

If f is predictable such that for any $T > 0$,

$$\mathbb{E}\int_0^T \|f(t)\|^2_{L^2(U,\mu)} dt < \infty,$$

it is easy to prove the convergence of the series

$$\sum_{j=1}^{\infty} \int_0^t \langle f(s), \phi^j\rangle_{L^2(U,\mu)} dW_s^j,$$

and the limit does not depend on the choice of the CONS. We abuse the notation a bit by omitting the variable ω and write f as $f(t,u)$. We denote the limit of the series above as the stochastic integral

$$M_t \equiv \int_0^t \int_U f(s,u) W(dsdu).$$

It is easy to show that M_t is a continuous martingale with quadratic variation process

$$\langle M \rangle_t = \int_0^t \int_U f(s,u)^2 \mu(du) ds.$$

We shall also need the definition of cylindrical Brownian motions, and the stochastic integration with respect to them.

Definition A.1.4. Let H be a separable Hilbert space with norm $\|\cdot\|_H$. A family $\{B_t(h) : t \geq 0, h \in H\}$ of real-valued random variables is called a *cylindrical Brownian motion (c.B.m.) on H with covariance* Σ if Σ is a continuous self-adjoint positive definite operator on H such that the following conditions hold:
i) For each $h \in H$ such that $h \neq 0$, $\langle \Sigma h, h \rangle_H^{-1/2} B_t(h)$ is a one-dimensional standard Wiener process.
ii) For any $t \geq 0$, $\alpha_1, \alpha_2 \in \mathbb{R}$ and $f_1, f_2 \in H$

$$B_t(\alpha_1 f_1 + \alpha_2 f_2) = \alpha_1 B_t(f_1) + \alpha_2 B_t(f_2) \quad \text{a.s.}$$

iii) For each $h \in H$, $\{B_t(h)\}$ is an $\mathcal{F}_t^B$-martingale, where

$$\mathcal{F}_t^B = \sigma\{B_s(k) : s \leq t, k \in H\}.$$

$\{B_t(h) : t \geq 0, h \in H\}$ is called a standard H-c.B.m. or simply, H-c.B.m. if it is a H-c.B.m. with covariance $\Sigma = I$.

Let $f : \mathbb{R}_+ \times \Omega \to H$ be a predictable process satisfying

$$\mathbb{E} \int_0^T \|f(t)\|_H^2 dt < \infty, \qquad \forall\, T > 0.$$

Define the stochastic integral

$$I_t(f) \equiv \int_0^t \langle f(s), dB_s \rangle_H = \sum_{j=1}^\infty \int_0^t \langle f(s), h_j \rangle_H \, dB_s(h_j), \tag{A.1.2}$$

where $\{h_j\}$ is a CONS of H.

It is easy to prove the convergence of (A.1.2) and the limit does not depend on the choice of the CONS. Furthermore, $I_t(f)$ is a square-integrable martingale with quadratic variation process

$$\langle I(f)\rangle_t = \int_0^t \|f(s)\|_H^2 ds.$$

We now present a martingale representation theorem of Kallianpur and Xiong (1995) (we refer the reader to Theorem 3.3.6 there for details). Note that $\mathcal{S}(\mathbb{R})$, the space of rapidly decreasing functions, is a nuclear space. We shall state the theorem in terms of $\mathcal{S}(\mathbb{R})$ instead of the general nuclear space. The dual $\mathcal{S}'(\mathbb{R})$, Schwartz distribution space, is a conuclear space in the terminology of Kallianpur and Xiong (1995).

Definition A.1.5. An $\mathcal{S}'(\mathbb{R})$-valued process $M = \{M_t\}_{t\geq 0}$ is an $\mathcal{S}'(\mathbb{R})$-*martingale* with respect to $\{\mathcal{F}_t\}$ if for each $\phi \in \mathcal{S}(\mathbb{R})$, $M_t(\phi)$ is a martingale with respect to $\{\mathcal{F}_t\}$. It is called an $\mathcal{S}'(\mathbb{R})$-*square-integrable-martingale* if, in addition,

$$E(M_t(\phi)^2) < \infty, \qquad \forall \phi \in \mathcal{S}(\mathbb{R}),\ t \geq 0.$$

We denote the collection of $\mathcal{S}'(\mathbb{R})$-martingales (resp. $\mathcal{S}'(\mathbb{R})$-square-integrable-martingales) by $\mathcal{M}(\mathcal{S}(\mathbb{R})')$ (resp. $\mathcal{M}^2(\mathcal{S}(\mathbb{R})')$). We also denote

$$\mathcal{M}^{2,c}(\mathcal{S}(\mathbb{R})') = \left\{ M \in \mathcal{M}^2(\mathcal{S}(\mathbb{R})') : \quad \begin{array}{l} M_t(\phi) \text{ has a continuous} \\ \text{version for each } \phi \in \mathcal{S}(\mathbb{R}) \end{array} \right\}.$$

Now we introduce the concept of $\mathcal{S}'(\mathbb{R})$-valued Wiener process.

Definition A.1.6. A continuous $\mathcal{S}'(\mathbb{R})$-valued stochastic process $W = (W_t)_{t\geq 0}$ on $(\Omega, \mathcal{F}, P)$ is called a centered $\mathcal{S}'(\mathbb{R})$-*Wiener process with covariance* $Q(\cdot,\cdot)$ if W satisfies the following three conditions:
a) $W_0 = 0$ a.s.
b) W has independent increments, i.e. random variables

$$W_{t_1}(\phi_1), (W_{t_2} - W_{t_1})(\phi_2), \cdots, (W_{t_n} - W_{t_{n-1}})(\phi_n)$$

are independent for any $\phi_1, \cdots, \phi_n \in \mathcal{S}(\mathbb{R})$, $0 \leq t_1 \leq \cdots \leq t_n, n \geq 1$.
c) For each $t \geq 0$ and $\phi \in \mathcal{S}(\mathbb{R})$

$$E\left(e^{iW_t(\phi)}\right) = e^{-tQ(\phi,\phi)/2}$$

where Q is a *covariance functional*, i.e. a positive definite symmetric continuous bilinear form on $\mathcal{S}(\mathbb{R}) \times \mathcal{S}(\mathbb{R})$.

Note that $\mathcal{S}'(\mathbb{R})$-valued Wiener process W can be regarded as an H_Q-c.B.m. if we define H_Q as the Hilbert space obtained by the completion of $\mathcal{S}(\mathbb{R})$ with respect to the norm

$$\|\phi\|_{H_Q} = Q(\phi, \phi)^{1/2}.$$

Denote the collection of all continuous linear operators from space $\mathcal{X}$ to space $\mathcal{Y}$ by $L(\mathcal{X}, \mathcal{Y})$. Let f be a predictable $L(\mathcal{S}'(\mathbb{R}), \mathcal{S}'(\mathbb{R}))$-valued process. Let $\{h_j\} \subset \mathcal{S}(\mathbb{R})$ be a CONS of H_Q. We define the stochastic integral

$$M_t = \int_0^t f(s)dW_s$$

as an $\mathcal{S}'(\mathbb{R})$-valued martingale as follows: For all $\phi \in \mathcal{S}(\mathbb{R})$,

$$\langle M_t, \phi\rangle = \sum_{j=1}^{\infty} \int_0^t \langle f(s)^*\phi, h_j\rangle_{H_Q} dW_s(h_j),$$

where $f(s)^* \in L(\mathcal{S}(\mathbb{R}), \mathcal{S}(\mathbb{R}))$ is the adjoint operator of $f(s)$.

We now state the martingale representation theorem without proof.

Theorem A.1.7. *Let Q be a covariance function on $\mathcal{S}(\mathbb{R}) \times \mathcal{S}(\mathbb{R})$. Suppose that $M \in \mathcal{M}^{2,c}(\mathcal{S}'(\mathbb{R}))$ and there exists a predictable $L(\mathcal{S}'(\mathbb{R}), \mathcal{S}'(\mathbb{R}))$-valued process f such that*

$$\langle M(\phi)\rangle_t = \int_0^t Q(f(s)^*\phi, f(s)^*\phi)ds.$$

Then on an extension $(\tilde{\Omega}, \tilde{\mathcal{F}}, \tilde{P}, \tilde{\mathcal{F}}_t)$ of $(\Omega, \mathcal{F}, P, \mathcal{F}_t)$, there exists a $\mathcal{S}'(\mathbb{R})$-Wiener process W with covariance Q such that

$$M_t = \int_0^t f(s)dW_s. \tag{A.1.3}$$

Theorem A.1.7 can be generalized as follows.

Theorem A.1.8. *Let Q be a covariance function on $\mathcal{S}(\mathbb{R}) \times \mathcal{S}(\mathbb{R})$ and H be a Hilbert space. Suppose that $M \in \mathcal{M}^{2,c}(\mathcal{S}'(\mathbb{R}))$ and there exist a predictable $L(\mathcal{S}'(\mathbb{R}), \mathcal{S}'(\mathbb{R}))$-valued process f and a predictable $L(\mathcal{S}(\mathbb{R}), H)$-valued process g such that*

$$\langle M_t(\phi)\rangle = \int_0^t \left(Q(f(s)^*\phi, f(s)^*\phi) + \|g(s)\phi\|_H^2\right) ds.$$

Then on an extension $(\tilde{\Omega}, \tilde{\mathcal{F}}, \tilde{P}, \tilde{\mathcal{F}}_t)$ of $(\Omega, \mathcal{F}, P, \mathcal{F}_t)$, there exist a $\mathcal{S}'(\mathbb{R})$-Wiener process W with covariance Q and an H-c.B.m. B, which is independent of W, such that

$$M_t = \int_0^t f(s)dW_s + \int_0^t g(s)^*dB_s. \tag{A.1.4}$$

Finally, we demonstrate that a stochastic integral with respect to an H-c.B.m. can be converted to one with respect to a white noise random measure when $H = L^2(U, \mu)$. Suppose that B_t is an H-c.B.m. and

$$M_t = \int_0^t \langle f(s), dB_s \rangle_H ,$$

where f is an H-valued predictable process. Let $\{h_j\}$ be a CONS of H. We define the white noise random measure W on $\mathbb{R}_+ \times U$ as

$$W([0,t] \times A) = \sum_{j=1}^{\infty} \langle 1_A, h_j \rangle_H B_t(h_j), \qquad \forall\, A \in \mathcal{U}.$$

Then,

$$M_t = \int_0^t \int_U f(s,u) W(dsdu).$$

A.2 Weak convergence

In this section, we state a result of Kotelenez and Kurtz (2010) about weak convergence of the empirical measure process for infinite particle system. It has been used to establish the existence of solutions for infinite system of SDEs. Let E be a complete separable metric space. The notation $X_n \Rightarrow X$ stands for the convergence of X_n to X in laws.

Theorem A.2.1. *For $n = 1, 2, \cdots$, let $X^n = (X^{1,n}, \cdots, X^{N_n,n})$ be exchangeable families of $D(\mathbb{R}_+, E)$-valued random variables such that $N_n \Rightarrow \infty$ and $X^n \Rightarrow X$ in $D(\mathbb{R}_+, E)^\infty$. Define*

$$\Xi_n = \frac{1}{N_n} \sum_{i=1}^{N_n} \delta_{X^{i,n}} \in \mathcal{P}(D(\mathbb{R}_+, E)),$$

$$\Xi = \lim_{m \to \infty} \frac{1}{m} \sum_{i=1}^{m} \delta_{X^i},$$

$$Z_t^n = \frac{1}{N_n} \sum_{i=1}^{N_n} \delta_{X_t^{i,n}} \in \mathcal{P}(E),$$

and

$$Z_t = \lim_{m \to \infty} \frac{1}{m} \sum_{i=1}^{m} \delta_{X_t^i},$$

and set

$$D_\Xi = \{t : E[\Xi\{x : x_t \neq x_{t-}\}] > 0\}.$$

Then the following hold.

a) For $t_1, \cdots, t_l \notin D_\Xi$,

$$(\Xi_n, Z^n_{t_1}, \cdots, Z^n_{t_l}) \Rightarrow (\Xi, Z_{t_1}, \cdots, Z_{t_l}).$$

b) If $X^n \Rightarrow X$ *in* $D(\mathbb{R}_+, E^\infty)$, *then*

$$(X^n, Z^n) \Rightarrow (X, Z) \quad \text{in } D(\mathbb{R}_+, E^\infty \times \mathcal{P}(E)).$$

If $X^n \to X$ *in probability in* $D(\mathbb{R}_+, E^\infty)$, *then* $(X^n, Z^n) \to (X, Z)$ *in* $D(\mathbb{R}_+, E^\infty \times \mathcal{P}(E))$ *in probability.*

A.3 Relation among strong existence, weak existence and pathwise uniqueness

For finite dimensional Itô equations, Yamada and Watanabe (1971) proved that weak existence and strong uniqueness imply strong existence and weak uniqueness. Kurtz (2007) considered this problem in an abstract setting which we outline below.

Let S_1 and S_2 be Polish spaces, and let $\Gamma: S_1 \times S_2 \to \mathbb{R}$ be a Borel measurable function. Let Y be an S_2-valued random variable with distribution ν. We are interested in solutions of the equation

$$\Gamma(X, Y) = 0. \tag{A.3.1}$$

Let $\mathcal{S}_{\Gamma,\nu}$ be the collection of $\mu \in \mathcal{P}(S_1 \times S_2)$ such that $\mu = \mathcal{L}(X, Y)$, the distribution of (X, Y) which solves (A.3.1) and $\mu(S_1 \times \cdot) = \nu$. Each $\mu \in \mathcal{S}_{\Gamma,\nu}$ is called a joint solution measure.

A strong solution is a measurable mapping $F: S_2 \to S_1$ such that for any S_2-valued random variable Y with distribution ν, $(F(X), Y)$ is a solution to (A.3.1).

Definition A.3.1. Pointwise uniqueness holds for (A.3.1) if X_1, X_2 and Y defined on the same probability space with $\mathcal{L}(X_1, Y) = \mathcal{L}(X_2, Y) \in \mathcal{S}_{\Gamma,\nu}$ implies $X_1 = X_2$ a.s.

For $\mu \in \mathcal{S}_{\Gamma,\nu}$, μ-uniqueness holds if X_1, X_2 and Y defined on the same probability space with $\mathcal{L}(X_1, Y) = \mathcal{L}(X_2, Y) = \mu$ implies $X_1 = X_2$ a.s.

Uniqueness in law (or weak uniqueness) holds if all $\mu \in \mathcal{S}_{\Gamma,\nu}$ have the same marginal distribution on S_1.

Theorem A.3.2. *If* $\mu \in \mathcal{S}_{\Gamma,\nu}$ *and* μ*-uniqueness holds, then* μ *is the joint distribution for a strong solution.*

Now, we give the equivalence of various uniqueness properties.

Theorem A.3.3. *The following are equivalent:*

(1) Pointwise uniqueness.
(2) μ-uniqueness for every $\mu \in \mathcal{S}_{\Gamma,\nu}$.
(3) Uniqueness in law.

Sometimes, (X, Y) must satisfies some extra conditions. We refer such conditions as a *compatible structure*, denoted by $\mathcal{C}$. For any $\nu \in \mathcal{P}(S_2)$, let $\mathcal{S}_{\mathcal{C},\nu}$ be the collection of $\mu \in \mathcal{P}(S_1 \times S_2)$ with the following properties:

(1) $\mu(S_1 \times \cdot) = \nu$,
(2) If (X, Y) has distribution μ, then X is compatible with Y.

It can be proved that $\mathcal{S}_{\mathcal{C},\nu}$ is a convex subset of $\mathcal{P}(S_1 \times S_2)$.

Let Γ be a collection of constraints that determine convex subsets of $\mathcal{P}(S_1 \times S_2)$, and $\mathcal{S}_{\Gamma,\mathcal{C},\nu}$ be the collection of $\mu \in \mathcal{S}_{\mathcal{C},\nu}$ such that μ fulfills the constraints in Γ.

Theorem A.3.4. *Suppose that $\mathcal{S}_{\Gamma,\mathcal{C},\nu} \neq \emptyset$. The following are equivalent:*

(1) Pointwise uniqueness holds for compatible solutions.
(2) Joint uniqueness in law holds for compatible solutions and there exists a strong, compatible solution.

The most useful case is for an SDE with Y as the driving and X as the solution. In this case, X and Y are cádlǵ processes taking values in E_1 and E_2), respectively. X is compatible with Y if for any $t \geq 0$ and h a bounded measurable function on $D(\mathbb{R}_+, E_2)$, we have

$$\mathbb{E}\left(h(Y)|\mathcal{F}_t^{X,Y}\right) = \mathbb{E}\left(h(Y)|\mathcal{F}_t^{Y}\right), \qquad \forall t \geq 0.$$

When the SDE is written as

$$X_t = x_0 + \int_0^t H(X, s-)dY_s,$$

with Y_s a semimartingale, the constrains usually consists: For any $t > 0$,

$$\lim_{n\to\infty} \mathbb{E}\left(1 \wedge \left|X_t - X_0 - \sum_{k=0}^{n-1} H(X, n^{-1}k)\left(Y_{(k+1)n^{-1}\wedge t} - Y_{kn^{-1}\wedge t}\right)\right|\right) = 0.$$

Bibliography

Adams, R. A. (1975). *Sobolev Spaces.* Pure and Applied Mathematics, Vol. 65. Academic Press, New York-London.

Bain, A. and Crisan, D. (2009). *Foundamentals of Stochastic Filtering.* Stochastic Modeling and Applied Probability, **60**. Springer, New York.

Barros-Neto, J. (1973). *An Introduction to the Theory of Distributions.* Marcel Dekker, Inc. New York.

Bernard, P., Talay, D. and Tubaro, L. (1994). Rate of convergence for the Kolmogorov equation with variable coefficients. *Math. Comp.* **63**, 555-587.

Billingsley, P. (1979). *Probability and Measure.* John Wiley and Sons, New York.

Bismut, J. M. (1978). An introductory approach to duality in optimal stochastic control. *SIAM Rev.* **20**, 62-78.

Burdzy, C., Mueller, C. and Perkins, E. A. (2012). Non-uniqueness for non-negative solutions of parabolic SPDE's, *Ill. J. Math.*, To appear.

Carmona, R. and Viens, F. (1998). Almost-sure exponential behavior of a stochastic Anderson model with continuous space parameter. *Stochastics Stochastics Rep.* **3-4**, 251-273.

Chiang, T., Kallianpur, G., and Sundar, P. (1991). Propagation of chaos and McKean-Vlasov equation in duals of nuclear spaces, *Applied Mathematics and Optimization* **24**, 55-83.

Crisan, D. (2003). Superprocesses in a Brownian environment. Stochastic analysis with applications to mathematical finance. *Proc. R. Soc. Lond. Ser. A Math. Phys. Eng. Sci.* **460**, no. 2041, 243–270.

Crisan, D., Gaines, J. and Lyons, T. (1998). Convergence of a branching particle method to the solution of the Zakai equation. *SIAM J. Appl. Math* **58**, No. 5, 1568–1590.

Crisan, D. and Lyons, T. (1997). Nonlinear filtering and measure-valued processes. *Probab. Theory Related Fields,* **109**, 217-244.

Crisan, D. and Xiong, J. (2007). A central limit type theorem for a class of particle filters. *Commun. Stoch. Anal.* **1**, no. 1, 103122.

Da Prato, G. and Tubaro, L. eds. (2002). Stochastic Partial Differential Equations and Applications. Lecture Notes in Pure and Appl. Math. 227. Springer, Berlin.

Da Prato, G. and Zabczyk, J. (1992). *Stochastic equations in infinite dimensions.* Encyclopedia of Mathematics and its Applications, 44. Cambridge University Press, Cambridge.

Dawson, D. A. (1975). Stochastic evolution equations and related measure processes. *J. Multivariate Anal.* **5**, 152.

Dawson, D. A. (1993). Measure-valued Markov processes. École d'Été de Probabilités de Saint-Flour XXI—1991, 1–260, *Lecture Notes in Math.,* **1541**, Springer, Berlin.

Dawson, D. A., Etheridge, A., Fleischmann, K., Mytnik, L., Perkins, E. A. and Xiong, J. (2002). Mutually catalytic processes in the plane: Infinite measure states, *Electronic Journal of Probability,* **7**, Paper no. 15, pages 1-61.

Dawson, D. A. and Feng, S. (1998). Large deviations for the Fleming-Viot process with neutral mutation and selection. *Stochastic Process. Appl.* **77**, no. 2, 207-232.

Dawson, D. A. and Feng, S. (2001). Large deviations for the Fleming-Viot process with neutral mutation and selection. II. *Stochastic Process. Appl.* **92**, no. 1, 131-162.

Dawson, D. A. and Li, Z. H. (2012). Stochastic equations, flows and measure-valued processes. *Ann. Probab.* **40**, no. 2, 813-857.

Dawson, D. A., Li, Z. H. and Wang, H. (2001). Superprocesses with dependent spatial motion and general branching densities. *Electron. J. Probab.* **6**, 1-33.

Dawson, D. A. and Vaillancourt, J. (1995). Stochastic McKean-Vlasov equations. *NoDEA Nonlinear Differential Equations Appl.* **2**, 199–229.

Dawson, D. A., Vaillancourt, J. and Wang, H. (2000). Stochastic partial differential equations for a class of interacting measure-valued diffusions. *Ann. Inst. Henri Poincaré Probab. Stat.* **36**, no. 2, 167-180.

Del Moral, P. (1996). Non-linear filtering: interacting particle resolution. *Markov Process. Related Fields* **2**, no. 4, 555-581.

Del Moral, P. (1995). Non-linear filtering using random particles. *Theory Probab. Appl.* **40**, 690-701.

Duncan, T. E. (1967). Doctoral Dissertation, Department of Electrical Engineering, Stanford University.

Dynkin, E. B. (1994). *An Introduction to Branching Measure-Valued Processes. CRM Monograph Series,* **6**. American Mathematical Society, Providence, RI.

Dynkin, E. B. (2002). *Diffusions, Superdiffusions and Partial Differential Equations.* American Mathematical Society Colloquium Publications, 50. American Mathematical Society, Providence, RI.

Dynkin, E. B. (1993). Superprocesses and partial differential equations. *Ann. Probab.* **21**, 1185-1262.

Etheridge, A. M. (2000). *An Introduction to Superprocesses.* University Lecture Series 20. American Mathematical Society.

Ethier, S. N. and Kurtz, T. G. (1986). *Markov Processes: Characterization and Convergence*, Wiley.

Fatheddin, P. and Xiong, J. (2012). Large deviation principle for some measure-valued processes. *Submitted.*

Feng, S. and Xiong, J. (2002). Large deviations and quasi-potential of a Fleming-Viot process. *Electron. Comm. Probab.* **7**, 13-25.

Florchinger, P. and Le Gland, F. (1992). Particle approximations for first order stochastic partial differential equations. *Applied stochastic analysis (New Brunswick, NJ,* 1991), 121-133, *Lecture Notes in Control and Inform. Sci.*, **177** Springer, Berlin.

Foondum, M. and Khoshnevisan, D. (2009). Intermittence and nonlinear parabolic stochastic partial differential equations. *Electron. J. Probab.* **14**, 548–568.

Friedman, A. (1975). Stochastic Differential Equations and Applications 1. Academic Press, New York.

Fujisaki, M., Kallianpur, G. and Kunita, H. (1972). Stochastic differential equations for the non-linear filtering problem. *Osaka J. Math.* **9**, 19-40.

Gel'fand, I. M. and Shilov, G. E. (1964). *Generalized functions. Vol. 1. Properties and operations.* Translated from the Russian by Eugene Saletan. Academic Press, New York-London.

Gomez, A., Lee, K., Mueller, C., Wei, A. and Xiong, J. (2012). Strong Uniqueness for an SPDE via backward doubly stochastic differential equations. *Work in progress.*

Graham, C. (1992). Nonlinear Ito-Skorohod equations and martingale problem with discrete jump sets, *Stochastic Process. Appl.* **40**, 69-82.

Gyöngy, I. (2002). Approximations of stochastic partial differential equations. Stochastic Partial Differential Equations and Applications. *Lecture Notes in Pure and Appl. Math.* **227**, 287-307. Springer, Berlin.

Hitsuda, M. and Mitoma, I. (1986). Tightness problem and stochastic evolution equation arising from fluctuation phenomena for interacting diffusions, *Journal of Multivariate Analysis* **19**, 311-328.

Hu, Y. Z., Lu, F. and Nualart, D. (2012). Hölder continuity of the solution for a class of nonlinear SPDEs arising from one-dimensional superprocesses. To appear in *Probability Theory Related Fields.*

Ibragimov, I. A. (1983). On smoothness conditions for trajectories of random functions. *Theory Probab. Appl.* **28**, 240-262.

Ikeda, N. and Watanabe, S. (1989). *Stochastic differential equations and diffusion processes.* Second edition. North-Holland Mathematical Library, 24. North-Holland Publishing Co., Amsterdam; Kodansha, Ltd., Tokyo.

Jacod, J. and Shiryaev, A. N. (1987). *Limit Theorems for Stochastic Processes.* Springer-Verlag.

Jirina, M. (1958). Stochastic branching processes with continuous state space. *Czech. Math. J.* **8**, 292-313.

Jirina, M. (1964). Branching processes with measure-valued states. In: *Trans. Third Prague Conf. Info. Th., Statis. Decision Functions, Random Processes*, 333-357. Publ. House Czech. Acad. Sci., Prague.

Kallenberg, O. (2002). *Foundations of modern probability. Second edition.* Probability and its Applications (New York). Springer-Verlag, New York.

Kallianpur, G. (1980). *Stochastic Filtering Theory.* Springer-Verlag.

Kallianpur, G. and Xiong, J. (1995). *Stochastic Differential Equations in Infinite*

Dimensional Spaces. IMS Lecture Notes - Monograph Series 26.

Kallianpur, G. and Xiong, J. (1994). Asymptotic behavior of a system of interacting nuclear-space-valued stochastic differential equations driven by Poisson random measures, *Applied Mathematics and Optimization* **30**, 175-201.

Konno, N. and Shiga, T. (1988). Stochastic partial differential equations for some measure-valued diffusions. *Probab. Theory Related Fields* **79**, no. 2, 201–225.

Kotelenez, P. M. (1992). Existence, uniqueness and smoothness for a class of function valued stochastic partial differential equations. *Stochastics Stochastics Rep.* **41** 177-199.

Kotelenez, P. M. (1992). Comparison methods for a class of function valued stochastic partial differential equations. *Probab. Theory Relat. Fields* **93**, 1-19.

Kotelenez, P. M. (1995). A class of quasilinear stochastic partial differential equation of McKean-Vlasov type with mass conservation, *Probab. Theory Relat. Fields* **102**, 159-188.

Kotelenez, P. M. and Kurtz, T. G. (2010) Macroscopic limits for stochastic partial differential equations of McKean-Vlasov type. *Probab. Theory Related Fields* **146**, no. 1-2, 189-222.

Krasnoselskii, M. A., Pustylnik, E. I., Sobolevski, P. E. and Zabrejko, P. P. (1976). *Integral Operators in Spaces of Summable Functions*, Nauka, Moscow, 1966 in Russian; English translation: Noordhoff International Publishing, Leyden.

Krylov, N. V. (1999). An analytic approach to SPDEs, Stochastic partial differential equations: six perspectives, *Math. Surveys Monogr.* **64**, 185-242, Amer. Math. Soc., Providence, RI.

Kunita, H. (1990). *Stochastic flows and stochastic differential equations.* Cambridge University Press.

Kurtz, T. G. (2007). The Yamada-Watanabe-Engelbert theorem for general stochastic equations and inequalities. *Electron. J. Probab.* **12**, 951-965.

Kurtz, T. G. and Protter, P. (1996). Weak convergence of stochastic integrals and differential equations. II. Infinite-dimensional case. *Probabilistic models for nonlinear partial differential equations, Lecture Notes in Math.*, **1627**, 197-285.

Kurtz, T. G. and Xiong, J. (1999). Particle representations for a class of nonlinear SPDEs. *Stochastic Processes and their Applications* **83**, 103-126.

Kurtz, T. G. and Xiong, J. (2004). A stochastic evolution equation arising from the fluctuation of a class of interacting particle systems. *Commun. Math. Sci.* **2**, 325–358.

Kushner, H. J. (1967). Dynamic equations for nonlinear filtering. *J. Differ. Equations,* **3**, 179-190.

Lee, K. J., Mueller, C. and Xiong, J. (2009). Some properties for superprocess over a stochastic flow. *Ann. Inst. Henri Poincaré Probab. Stat.* **45**, 477-490.

Le Gall, J.-F. (1999). *Spatial branching processes, random snakes and partial differential equations.* Lectures in Mathematics ETH Zrich. Birkhäuser Verlag, Basel.

Li, Z. H. (2011). *Measure-valued branching Markov processes.* Probability and its Applications (New York). Springer, Heidelberg.

Li, Z. H.; Ma, C. H. (2008). Catalytic discrete state branching models and related limit theorems. *Journal of Theoretical Probability* **21**, 4: 936–965.

Li, Z. H., Wang, H. and Xiong, J. (2005). Conditional Log-Laplace Functionals of Superprocesses with Dependent Spatial Motion. *Acta Applicandae Mathematicae* **88**, 143-175.

Li, Z. H.; Wang, H.; Xiong, J. (2004). A degenerate stochastic partial differential equation for superprocesses with singular interaction. *Probability Theory and Related Fields* **130**, 1: 1–17.

Li, Z. H.; Wang, H.; Xiong, J. and Zhou, X. W. (2012), Joint continuity of the solutions to a class of nonlinear SPDEs. *Probab. Theory Relat. Fields* **153**, No. 3, 441-469.

McKean, H. P. (1967). Propagation of Chaos for a Class of Non-linear Parabolic Equations, *Lecture Series in Differential Equations* **2**, 177-194.

Méléard, S. (1996). Asymptotic behavior of some interacting particle systems, McKean-Vlasov and Boltzmann models, *Probabilistic models for nonlinear partial differential equations, Lecture Notes in Math.*, **1627**, 42-95.

Mitoma, I. (1985). An ∞-dimensional inhomogeneous Langevin equation. *J. Funct. Anal.* **61**, 342-359.

Morien, P. L. (1996). Propagation of chaos and fluctuations for a system of weakly interacting white noise driven parabolic SPDE's. *Stochastics Stochastics Reps.* **58**, 1-43.

Mueller, C., Mytnik, L. and Perkins, E. A. (2012). Non-uniqueness for parabolic stochastic partial differential equations with Holder continuous coefficients. Submitted.

Mortensen, R. (1966). Doctoral Dissertation, Department of Electrical Engineering, University of California, Berkeley.

Mytnik, L. (1996). Superprocesses in random environments. *Ann. Probab.* **24**, No. 4. 1953-1978.

Mytnik, L. and Perkins, E. A. (2011). Pathwise uniqueness for stochastic heat equations with Hölder continuous coefficients: the white noise case. *Probab. Theory Related Fields* **149**, no. 1-2, 1-96.

Mytnik, L., Perkins, E. A. and Sturm, A. (2006). On pathwise uniqueness for stochastic heat equations with non-Lipschitz coefficients. *Ann. Probab.* **34**, no. 5, 1910–1959.

Mytnik, L. and Xiong, J. (2012). Well-posedness of the martingale problem for superprocess with interaction. In preparation.

Mytnik, L. and Xiong, J. (2007). Local extinction for superprocesses in random environments. *Electron. J. Probab.* **12**, 1349-1378.

Pardoux, E. and Peng, S. (1990). Adapted solution of backward stochastic equation. *Syst. & Control Lett.* **14**, 55-61.

Pardoux, E. and Peng, S. (1994). Backward doubly stochastic differential equations and systems of quasilinear SPDEs. *Probab. Th. Related Fields* **98**, 209-227.

Perkins, E.A. (2002). *Dawson-Watanabe Superprocesses and Measure-Valued Dif-*

fusions. Ecole d'Eté de Probabiliyés de Saint-Flour XXIX-1999. Lecture Notes Math. **1781**, 125-329. Bernard, P. ed. Springer, Berlin.

Reimers, M. (1989). One-dimensional stochastic partial differential equations and the branching measure diffusion. *Probab. Theory Related Fields* **81**, no. 3, 319-340.

Rozovskii, B. L. (1990). Stochastic Evolution Systems. Linear Theory and Applications to Nonlinear Filtering. Kluwer, Dordrecht.

Skoulakis, G. and Adler, R. J. (2001). Superprocesses over a stochastic flow. *Ann. Appl. Probab.* **11**, 488-543.

Sowers, R. B. (1995). Intermittency-type estimates for some nondegenerate SPDE's. *Ann. Probab.* 23, 1853-1874.

Viot, M. (1976). Solutions faibles d'equations aux drivees partielles stochastique non lineaires. Ph.D. thesis, Univ. Pierre et Marie Curie-Paris VI.

Wang, H. (1997). State classification for a class of measure-valued branching diffusions in a Brownian medium. *Probab. Theory Related Fields* **109**, 39–55.

Wang, H. (1998). A class of measure-valued branching diffusions in a random medium. *Stochastic Anal. Appl.* **16**, 753-786

Watanabe, S. (1968). A limit theorem of branching processes and continuous state branching processes. *J. Math. Kyoto Univ.* **8**, 141-167.

Xiang, K. and Zhang, T. (2005). Small time asymptotics for Fleming-Viot processes. *Infin. Dimens. Anal. Quantum Probab. Relat. Top.* **8**, no. 4, 605630.

Xiong, J. (2008). *An introduction to stochastic filtering theory.* Oxford University Press.

Xiong, J. (1995). A nonlinear differential equation associated with a class of interacting diffusion systems. *Trends in Contemporary Infinite Dimensional Analysis and Quantunm Probability.* Edited by L. Accardi, H. H. Kuo, N. Obata, K. Saito, S. Si and L. Streit. 2000, 433-457.

Xiong, J. (2004a). A stochastic log-Laplace equation. *Ann. Probab.* **32**, 2362-2388.

Xiong, J. (2004b). Long-term behavior for SBM over a stochastic flow. *Electronic Communications in Probability* **9**, 36-52.

Xiong, J. and Zhou, X. (2004). Superprocess over a stochastic flow with superprocess catalyst. *Int. J. Pure Appl. Math.* **17**, 353–382.

Yamada, T. and Watanabe, S. (1971). On the uniqueness of solutions of stochastic differential equations. *J. Math. Kyoto Univ.* **11** 155-167.

Zakai, M. (1969). On the optimal filtering of diffusion processes. *Z. Wahr. Verw. Gebiete* **11**, 230-243.

Index